新工科视域下的
工程基础与应用研究

《天水师范学院60周年校庆文库》编委会 | 编

光明日报出版社

图书在版编目（CIP）数据

新工科视域下的工程基础与应用研究 /《天水师范学院 60 周年校庆文库》编委会编. --北京：光明日报出版社，2019.9

ISBN 978-7-5194-5506-4

Ⅰ.①新… Ⅱ.①天… Ⅲ.①工程技术—基础科学—研究 Ⅳ.①TB1

中国版本图书馆 CIP 数据核字（2019）第 189337 号

新工科视域下的工程基础与应用研究
XINGONGKE SHIYU XIA DE GONGCHENG JICHU YU YINGYONG YANJIU

编　　者：《天水师范学院 60 周年校庆文库》编委会	
责任编辑：郭玫君	责任校对：赵鸣鸣
封面设计：中联学林	责任印制：曹　诤

出版发行：光明日报出版社
地　　址：北京市西城区永安路 106 号，100050
电　　话：010-67017249（咨询）　63131930（邮购）
传　　真：010-67078227，67078255
网　　址：http://book.gmw.cn
E - mail：guomeijun@gmw.cn
法律顾问：北京德恒律师事务所龚柳方律师
印　　刷：三河市华东印刷有限公司
装　　订：三河市华东印刷有限公司
本书如有破损、缺页、装订错误，请与本社联系调换，电话：010-67019571

开　　本：170mm×240mm			
字　　数：332 千字		印　张：18.5	
版　　次：2019 年 9 月第 1 版		印　次：2019 年 9 月第 1 次印刷	
书　　号：ISBN 978-7-5194-5506-4			
定　　价：89.00 元			

版权所有　　翻印必究

《天水师范学院60周年校庆文库》编委会

主　任：李正元　安　涛
副主任：师平安　汪聚应　王旭林　李　淳
　　　　汪咏国　安建平　王文东　崔亚军
　　　　马　超
委　员：王三福　王廷璞　王宏波　王贵禄
　　　　尤晓妮　牛永江　朱　杰　刘新文
　　　　李旭明　李艳红　杨　帆　杨秦生
　　　　张跟东　陈于柱　贾利珠　郭昭第
　　　　董　忠
编　务：刘　勋　汪玉峰　赵玉祥　施海燕
　　　　赵百祥　杨　婷　包文娟　吕婉灵

总　序

　　春秋代序,岁月倥偬,弦歌不断,薪火相传。不知不觉,天水师范学院就走过了它60年风雨发展的道路,迎来了它的甲子华诞。为了庆贺这一重要历史时刻的到来,学校以"守正·奋进"为主题,筹办了缤纷多样的庆祝活动,其中"学术华章"主题活动,就是希冀通过系列科研活动和学术成就的介绍,建构学校作为一个地方高校的公共学术形象,从一个特殊的渠道,对学校进行深层次也更具力度的宣传。

　　《天水师范学院60周年校庆文库》(以下简称《文库》)是"学术华章"主题活动的一个重要构成。《文库》共分9卷,分别为《现代性视域下的中国语言文学研究》《"一带一路"视域下的西北史地研究》《"一带一路"视域下的政治经济研究》《"一带一路"视域下的教师教育研究》《"一带一路"视域下的体育艺术研究》《生态文明视域下的生物学研究》《分子科学视域下的化学前沿问题研究》《现代科学思维视域下的数理问题研究》《新工科视域下的工程基础与应用研究》。每卷收录各自学科领域代表性科研骨干的代表性论文若干,集中体现了师院学术的传承和创新。编撰之目的,不仅在于生动展示每一学科60年来学术发展的历史和教学改革的面向,而且也在于具体梳理每一学科与时俱进的学脉传统和特色优势,从而体现传承学术传统,发扬学术精神,展示学科建设和科学研究的成就,砥砺后学奋进的良苦用心。

　　《文库》所选文章,自然不足以代表学校科研成绩的全部,近千名教职员工,60年孜孜以求,几代师院学人的学术心血,区区九卷书稿300多篇文章,个中内容,岂能一一尽显?但仅就目前所成文稿观视,师院数十年科研的旧貌新颜、变化特色，也大体有了一个较为清晰的眉目。

首先，《文库》真实凸显了几十年天水师范学院学术发展的历史痕迹，为人们全面了解学校的发展提供了一种直观的印象。师院的发展，根基于一些基础老学科的实力，如中文、历史、数学、物理、生物等，所以翻阅《文库》文稿，可以看到这些学科及其专业辉煌的历史成绩。张鸿勋、雒江生、杨儒成、张德华……，一个一个闪光的名字，他们的努力，成就了天水师范学院科研的初始高峰。但是随着时代的发展和社会需求的变化，新的学科和专业不断增生，新的学术成果也便不断涌现，教育、政法、资环等新学院的创建自是不用特别说明，单是工程学科方面出现的信息工程、光电子工程、机械工程、土木工程等新学科日新月异的发展，就足以说明学校从一个单一的传统师范教育为特色的学校向一个兼及师范教育但逐日向高水平应用型大学过渡的生动历史。

其次，《文库》具体显示了不同历史阶段不同师院学人不同的学术追求。张鸿勋、雒江生一代人对于敦煌俗文学、对于《诗经》《尚书》等大学术对象的文献考订和文化阐释，显见了他们扎实的文献、文字和学术史基本功以及贯通古今、熔冶正反的大视野、大胸襟，而雍际春、郭昭第、呼丽萍、刘雁翔、王弋博等中青年学者，则紧扣地方经济社会发展做文章，彰显地域性学术的应用价值，于他人用力薄弱或不及处，或成就了一家之言，或把论文写在陇原大地，结出了累累果实，发挥了地方高校科学研究服务区域经济社会发展的功能。

再次，《文库》直观说明了不同学科特别是不同学人治学的不同特点。张鸿勋、雒江生等前辈学者，其所做的更多是个人学术，其长处是几十年如一日，埋首苦干，皓首穷经，将治学和修身融贯于一体，在学术的拓展之中同时也提升了自己的做人境界。但其不足之处则在于厕身僻地小校之内，单兵作战，若非有超人之志，持之以恒，广为求索，自是难以取得理想之成果。即以张、雒诸师为例，以其用心用力，原本当有远愈于今日之成绩和声名，但其诸多未竟之研究，因一人之逝或衰，往往成为绝学，思之令人不能不扼腕以叹。所幸他们之遗憾，后为国家科研大势和学校科研政策所改变，经雍际春、呼丽萍等人之中介，至如今各学科纷纷之新锐，变单兵作战为团队攻坚，借助于梯队建设之良好机制运行，使一人之学成一众之学，前有所行，后有所随，断不因以人之故废以方向之学。

还有，《文库》形象展示了学校几十年科研变化和发展的趋势。从汉语到外语，变单兵作战为团队攻坚，在不断于学校内部挖掘潜力、建立梯队的同时，学校的一些科研骨干如邢永忠、王弋博、令维军、李艳红、陈于柱等，也融入了更大和更高一级的学科团队，从而不仅使个人的研究因之而不断升级，而且也带动学校的科研和国内甚至国际尖端研究初步接轨，让学校的声誉因之得以不断走向更远也更高更强的区域。

当然，前后贯通，整体比较，缺点和不足也是非常明显的，譬如科研实力的不均衡，个别学科长期的缺乏领军人物和突出的成绩；譬如和老一代学人相比，新一代学人人文情怀的式微等。本《文库》的编撰因此还有另外的一重意旨，那就是立此存照，在纵向和横向的多面比较之中，知古鉴今，知不足而后进，让更多的老师因之获得清晰的方向和内在的力量，通过自己积极而坚实的努力，为学校科研奉献更多的成果，在区域经济和周边社会的发展中提供更多的智慧，赢得更多的话语权和尊重。

六十年风云今复始，千万里长征又一步。谨祈《文库》的编撰和发行，能引起更多人对天水师范学院的关注和推助，让天水师范学院的发展能够不断取得新的辉煌。

是为序。

李正元　安涛

2019 年 8 月 26 日

目 录
CONTENTS

基于脉冲耦合神经网络的图像 NMI 特征提取及检索方法 …………… 刘 勍 1

结合灰度熵变换 PCNN 小目标图像检测新方法 …………………… 刘 勍 14

一种基于超模糊熵 ULPCNN 图像自动分割新方法 ………………… 刘 勍 24

一种频率域提高 Radon 变换分辨率的方法 ……………………… 刘保童 35

非均匀 DFT 频谱泄漏抑制方法研究 …………………………… 刘保童 47

一种随机振动信号幅度、相位及零基线漂移分离方法
………………………………………… 张少刚 李芙蓉 赵小龙 59

Numerical study of the influence of applied voltage on the current balance factor
of single layer organic light-emitting diodes …………………………………
………………… Lu Feiping Liu xiaobin Xing yongzhong 67

Numerical model of tandem organic light-emitting diodes based on transition metal oxide
interconnector layer ……… Lu Feiping Peng Yingquan Xing Yongzhong 81

High efficiency, high energy, CEP – stabilized infrared optical parametric amplifier …
………… Weijun Ling Xiaotao Geng Shuyan Guo Zhiyi Wei F. Krausz D. Kim 103

$1.91\mu m$ passively continuous-wavemode-locked $Tm:LiLuF_4$ laser ……………
………… WeiJun Ling Tao Xia Zhong Dong LiangFang You YinYan Zuo
            Ke Li QinLiu FeiPing Lu XiaoLong Zhao 115

Electronic Structures of Anatase $(TiO_2)_{1-x}(TaON)_x$ Solid Solutions:
First – Principles Study ……………………………………………………
………… Wenqiang Dang Hungru Chen Naoto Umezawa Junying Zhang 124

基于 NSGA – II 算法的管道清灰机器人变径机构优化 … 罗海玉 张淑珍 143

一种新型并联太阳能跟踪机构研究 …………………………… 罗海玉 152

磁力电解复合抛光中带电粒子运动的分析 ………………………… 牛永江 160

工件定位基准的定义及其确定新探 …………………………… 牛永江 167

Static and dynamic characteristics modeling for CK61125 CNC lathe bed basing on FEM ……………………………………… Hongping Yang 170

基于分形几何与接触力学理论的结合面法向接触刚度计算模型 … 杨红平 180

基于回声状态网络的结合面特性参数建模 ……………………… 杨红平 192

Investigation of Tool Wear and Surface Roughness when Turning Titanium Alloy (Ti6Al4V) under Different Cooling and Lubrication Conditions …………………………………………………… Limin Shi 204

基于小波包能量曲率差的古木结构损伤识别 ………………………………
…………………………………………… 王 鑫 胡卫兵 孟昭博 213

地面交通激励下西安钟楼木结构的损伤识别 ………… 王 鑫 孟昭博 230

西安钟楼的地震响应分析 …………………………… 王 鑫 孟昭博 245

一种基于脉冲耦合神经网络的图像双边滤波算法 ……………… 刘 勍 255

一种基于改进PCNN噪声检测的两级脉冲噪声滤波算法 ……… 刘 勍 264

天水古民居的建筑艺术与文化内涵研究 ………………………… 南喜涛 274

后　记 ……………………………………………………………………… 283

基于脉冲耦合神经网络的图像 NMI 特征提取及检索方法

刘 勍*

为简单有效地提取图像重要特征信息从而更好地提高检索图像的精度,提出了一种基于脉冲耦合神经网络(PCNN)图像归一化转动惯量(NMI)特征提取及检索算法。首先利用改进简化 PCNN 模型相似神经元同步时空特性及指数衰降机制将图像分解为具有相关性的二值系列图像,然后提取反映原始图像目标形状、结构分布二值系列图像的一维 NMI 特征矢量信号,并将其应用在图像检索中;同时,考虑到二值系列图像间的相关性及不同图像间 NMI 序列值的差异性,引入了马氏距离结合 Pearson 积矩相关法的综合相似性度量方法。实验结果表明,所提算法对图像特征矢量序列具有良好抗几何畸变不变特性及对图像表述的唯一性,且具有较好的图像检索效果。

1 引言

图像检索是图像处理和计算机视觉中重要的研究领域,而基于内容的图像检索又是目前图像检索的主要方法和研究热点,其核心思想是表征出图像色彩、形状、纹理及轮廓等不同内容的重要特征用来作为图像索引,并由此计算要查询图像和目标图像的相似性[1]。其中基于图像颜色检索主要利用颜色直方图进行图像间的相似性判断[2~3]或运用统计方法提取有感知的相关颜色信息特征[4],但存在易丢失颜色空间分布信息、图像颜色量化中会造成误检现象以及检索时间加长等问题;而基于形状的检索由于要采用边缘提取、边缘细化及形状描述等一系列几何学或拓扑处理方法[5~9],因此形状特征图像检索中形状特征的提取和分析又显得比较复杂;基于纹理检索由于一般图像的纹理特征不太显著,而在检索中对

---

* 作者简介:刘勍(1970—),男,甘肃天水人,天水师范学院教授、博士,主要从事图像信号处理及人工神经网络研究。

检索图像或区域纹理的一致性要求较高[10~11],其适用范围较小。

为此,本文在对脉冲耦合神经网络(Pulse Coupled Neural Networks,PCNN)[12]改进与简化的基础上,把 PCNN 和归一化转动惯量(normalized moment of inertia,NMI)相结合,提出了一种基于 PCNN 图像 NMI 特征矢量提取与检索算法。该方法首先利用改进 PCNN 模型对图像进行系列二值处理,再提取其一维 NMI 不变特征序列信号,并将其应用在图像检索中,同时引入距离结合相关性综合相似性度量,最后通过实验验证了所提算法的有效性。

2　PCNN 改进简化模型及图像二值序列分解

PCNN 也称为第三代人工神经网络,是 Eckhorn 等人在猫、猴等动物大脑视觉皮层模型启发下提出的由若干个互连神经元构成的反馈型网络,构成 PCNN 的每一神经元由接收、调制和脉冲产生 3 部分组成,目前已被广泛地应用于图像平滑、分割、边缘检测及目标检索等图像处理领域[13~16],显示了其优越性。

传统 PCNN 模型的关键思想是非线性调制耦合和阈值指数衰变机制,其中非线性调制耦合是其核心,而其阈值虽然指数衰降但又反复变化,显然,这种变化不符合视觉系统对亮度响应的非线性要求,同时这种阈值变化规律致使处理后许多神经元激活周期或激活相位中滞留大量有用信息,并且延长了 PCNN 的处理时间,而其直接的二值输出序列图像却并不包含全部信息。为克服传统 PCNN 人工设置参数多、适应性能差以及处理时间长等缺点,本文在传统 PCNN 模型的基础上进行了简化与改进,简化改进 PCNN 模型的离散数学方程描述如下:

$$F_{ij}[n] = I_{ij} \tag{1}$$

$$L_{ij}[n] = V_L \sum_{kl} W_{ijkl} Y_{ij}[n-1] \tag{2}$$

$$U_{ij}[n] = F_{ij}[n](1 + \beta L_{ij}[n]) \tag{3}$$

$$Y_{ij}[n] = \begin{cases} 1, if\, U_{ij}[n] \geq T_{ij}[n] \\ 0, if\, U_{ij}[n] < T_{ij}[n] \end{cases} \tag{4}$$

$$T_{ij}[n] = \begin{cases} T_0 e^{-\alpha_T(n-1)}, if\, Y_{ij}[n-1] = 0 \\ V_T, if\, Y_{ij}[n-1] = 1 \end{cases} \tag{5}$$

其中 ij 下标为神经元的标号,n 为迭代次数,I_{ij}、$F_{ij}[n]$、$L_{ij}[n]$、$U_{ij}[n]$、$T_{ij}[n]$ 分别为神经元的外部刺激(图像像素构成矩阵中第 ij 个像素的灰度值)、第 ij 个神经元第 n 次反馈输入、链接输入、内部活动项和动态阈值,W 为链接权矩阵,β 为链接强度,V_L、T_0 为链接幅度常数和阈值幅度常数,T_0 一般自适应选取为待处理图像的最大灰度值 I^{max},即 $T_0 = I^{max}$,V_T 为一设定的较大常数,α_T 为相应的衰减系数,$Y_{ij}[n]$是 PCNN 的二值输出。其中式(1)反馈输入 $F_{ij}[n]$和式(2)链接输入 $L_{ij}[n]$是在原模型基础上做了简化,而式(5)动态阈值 $T_{ij}[n]$是对其阈值的改进。

改进简化 PCNN 的工作原理是:在图像处理过程中,首先将一个 2 维改进型 PCNN 网络的 $M×N$ 个神经元分别与 2 维输入图像的 $M×N$ 个像素相对应,所有神经元结构相等且各个神经元的参数一致,在第 1 次迭代时,神经元的内部活动项就等于外部刺激 I_{ij},其初始阈值为 I^{max},若 I_{ij} 大于或等于初始阈值,这时神经元输出 $Y_{ij}[1]=1$,称为激活,此时其动态阈值 T_{ij}将急剧增大到 V_T 并一直保持不变,而其他未激活神经元($Y_{ij}[1]=0$)的动态阈值在其后处理中随时间(或迭代次数 n)指数衰减,并且在此之后的各次迭代中,被激活的神经元通过与之相邻神经元的连接作用而激励捕获邻接神经元,若邻接神经元的内部活动项大于其动态阈值,则被捕获激活,否则不能捕获。显然,如果邻接神经元与前一次迭代激活神经元所对应的像素具有相似强度,则邻接神经元容易被捕获激活,反之则不能被捕获激活。

目前,生成二值序列图像的方法较多,有按灰度等间隔二值量化法[17]及位平面二值处理[18-19]等方法,但前者没有考虑序列图像中邻域像素及不同图像间的相关性,位平面法对图像分解后只是简单保留了高位平面信息,造成图像目标信息损失。而 PCNN 二值序列图像分解充分考虑视觉处理系统的特点,在每次迭代处理中,利用某一神经元激活空时特性来触发其邻域相似神经元的集体激活,生成神经元集群对应图像中具有相似性质的某一小目标区域,然后由所有不同相似小目标区域组成该次迭代的一幅二值分割图像,并且在不同迭代时刻将产生代表和反映原图像特征的不同二值图像,由此便形成一个二值序列图像,图 1 所示为改进 PCNN 处理二值序列图像中的部分二值图像。

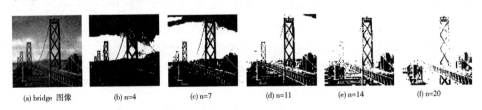

(a) bridge 图像　　(b) n=4　　(c) n=7　　(d) n=11　　(e) n=14　　(f) n=20

图 1　改进 PCNN 处理二值序列图像中迭代次数对应的部分二值图像

3　图像的特征描述

3.1　图像 NMI 特征

在物理学中将一个平面离散质点系的质心(在受均匀重力场作用时质心与重

心重合)表示为

$$x_c = \frac{\sum x_r m_r}{\sum m_r}, \quad y_c = \frac{\sum y_r m_r}{\sum m_r} \tag{6}$$

其中 m_r 为坐标 (x_r, y_r) 处质点的质量，$\sum m_r$ 为质点系的质量，(x_c, y_c) 为质点系的质心，它反应了质点系形状、大小及紧密程度等固有的一些特性。文献[20]将这一概念引入到数字图像处理领域，设2维图像大小为 $M \times N$ 个像素可看作是 XOY 平面上的 $M \times N$ 个质点，像素灰度值 I_{ij} 与相应质点的质量相对应，则对图像可做如下定义：

图像总质量：2维灰度图像所有的灰度值之和，记为 m，表示为

$$m = \sum_{i=1}^{M}\sum_{j=1}^{N} I_{ij} \tag{7}$$

图像重心：视为图像平面图像总质量集中的点，记为 (i_c, j_c)，可表示为

$$i_c = \frac{\sum_{i=1}^{M}\sum_{j=1}^{N} i I_{ij}}{\sum_{i=1}^{M}\sum_{j=1}^{N} I_{ij}}, \quad j_c = \frac{\sum_{i=1}^{M}\sum_{j=1}^{N} j I_{ij}}{\sum_{i=1}^{M}\sum_{j=1}^{N} I_{ij}} \tag{8}$$

图像的转动惯量：图像绕其中任一给定点 (i_0, j_0) 的转动惯量记为 $J_{i_0 j_0}$，表示为

$$J_{i_0 j_0} = \sum_{i=1}^{M}\sum_{j=1}^{N} [(i-i_0)^2 + (j-j_0)^2] I_{ij} \tag{9}$$

图像转动惯量与图像中不同目标的形状大小、灰度分布和转轴点的位置有关，但对灰度(或彩色)图像而言，由于其灰度分布比较复杂，不管转轴点选在图像重心或其他任何位置，其转动惯量都是几何(旋转、平移及缩放)畸变的，而二值图像只有0和1两种取值，其转动惯量具有良好的抗几何畸变特性。为此根据对图像总质量、图像重心及转动惯量的描述可定义二值图像绕重心的归一化转动惯量，简称归一化转动惯量 NMI，这里用 λ 表示

$$\lambda = \frac{\sqrt{J_{i_c j_c}}}{m} = \frac{\sqrt{\sum_{i=1}^{M}\sum_{j=1}^{N}[(i-i_c)^2+(j-j_c)^2]Y_{ij}}}{\sum_{i=1}^{M}\sum_{j=1}^{N} Y_{ij}}$$

$$= \frac{\sqrt{\sum_{i,j \in \Omega}[(i-i_c)^2+(j-j_c)^2]}}{\sum_{i,j \in \Omega} Y_{ij}} \tag{10}$$

式(10)中，Y_{ij} 为二值图像，Ω 为二值图像中 $Y_{ij}=1$ 的区域，可以看出 NMI 特征值

λ 为二值图像质量绕其重心的转动惯量与其质量之比。对不同的二值图像,可提取不同的 NMI 特征,并且 NMI 相对于传统的图像不变性特征(如图像矩特征、同心圆特征、拓扑特征等)具有提取方便、计算量小的特点。

3.2 二值序列图像 NMI 特征提取

视觉心理学家指出,人眼对一幅图像的观察认识是一个逐步剥离背景、集中于目标和部分重要细节的过程。受到这一过程的启发,本文利用符合人类视觉处理系统的 PCNN 模型,对待处理图像二值化后产生一系列彼此相关的二值图像,然后提取该二值系列图像的 NMI 不变性特征矢量序列。

利用改进 PCNN 模型在确定迭代次数 n_0 的情况下对任意图像 I_{ij} 运用式(1)~(5)进行逐层二值化处理,从而形成一个二值序列图像 $Y = \{Y[n], n = 1, 2\cdots, n_0\}$(图 1 为二值序列图像的部分图像),再利用式(10)分别计算序列图像中每幅图像的 NMI 值,最后得到该图像的一个 NMI 特征矢量 $\Lambda = \{\lambda_n, n = 1, 2, \cdots, n_0\}$。

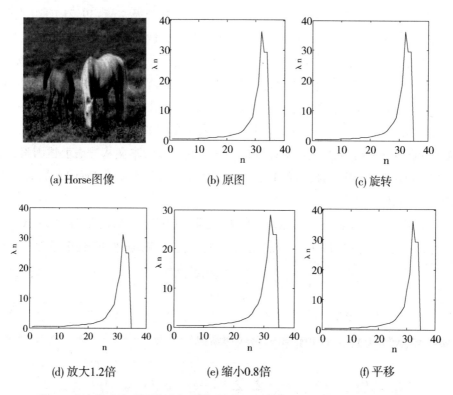

(a) Horse 图像　　(b) 原图　　(c) 旋转

(d) 放大 1.2 倍　　(e) 缩小 0.8 倍　　(f) 平移

图 2　本文算法中对图像进行旋转、缩放及平移提取的 NMI 特征矢量序列

图 2 所示为提取一幅图像的 NMI 特征矢量,它和单幅图像的 NMI 特征一样

具有旋转不变、尺度不变和平移不变性,对图像几何形变具有很强的鲁棒性。另外,二值图像序列中图像之间虽有一定的差异性,但又有很大的相关性,由此提取的各 NMI 特征值能客观反映原图像像素的空间分布、目标形状、结构大小等信息,同时其特征序列能够唯一代表一幅灰度图像特征。而且二值序列图像的 NMI 特征矢量计算复杂度较低且维数较少,可以满足图像检索中存储容量和检索速度的要求。

4 相似性度量

图像入库时,提取其 NMI 特征矢量放入图像特征信息库;检索时,提取查询图像的 NMI 特征矢量与图像特征信息库中的进行相似性比较,根据比较结果输出检索结果。在本文检索算法中,由于运用 PCNN 相似神经元同步时空特性产生某一图像二值序列,其中各图像之间有很强的相关性,使提取的 NMI 特征矢量间也存在较强联系,如果简单地用欧氏距离来求相似度,可能会造成相似图像 NMI 特征距离相差很大。而文献[17]采用 Pearson 积矩相关法以及文献[19,21]采用马氏距离(Mahalanobis distance)来度量图像间的相似性,虽然取得了一定的检索效果,但不同图像内容结构的复杂性和灰度的随机性,造成提取的 NMI 特征矢量总体不一定都成正态分布或接近正态单峰分布,而且检索的两矢量之间不一定成线性关系,这样 Pearson 积矩相关检索会引起较大的误检和漏检,另外,虽然马氏距离对检索数据的分布没有严格要求,也可排除检索矢量间的相互干扰,但会夸大变化微小数据量的作用,同样会影响图像的检索效果,为此,本文从考虑上述因素,在 PCNN 处理提取 NMI 特征检索的基础上,引入了马氏距离结合 Pearson 积矩相关法的综合相似性度量。

在 R^n 空间中,设两幅图像 I^A 和 I^B 提取的 NMI 特征向量分别为 $\Lambda^A = \{\lambda_n^A, n = 1,2,\cdots,n_0\}$ 和 $\Lambda^B = \{\lambda_n^B, n = 1,2,\cdots,n_0\}$,则综合相似性度量为

$$C(I^A, I^B) = \frac{D_M(\Lambda^A, \Lambda^B)}{|Corr(\Lambda^A, \Lambda^B)|} \tag{11}$$

当 $C(I^A, I^B)$ 越小,表示两幅图像的相似性越强。其中 $D_M(\Lambda^A, \Lambda^B)$ 为马氏距离, $Corr(\Lambda^A, \Lambda^B)$ 为 Pearson 积矩相关,分别表示为

$$D_M(\Lambda^A, \Lambda^B) = \sqrt{\sum_{u=1}^{n_0} \sum_{v=1}^{n_0} w_{uv} (\lambda_u^A - \lambda_u^B)(\lambda_v^A - \lambda_v^B)} \tag{12}$$

用向量表示形式是

$$D_M(\Lambda^A, \Lambda^B) = \sqrt{(\Lambda^A - \Lambda^B)^T S^{-1} (\Lambda^A - \Lambda^B)} \tag{13}$$

$$Corr(\Lambda^A, \Lambda^B) = \frac{\sum_{u=v=1}^{n_0}(\lambda_u^A - \overline{\Lambda^A})(\lambda_v^B - \overline{\Lambda^B})}{\sqrt{\sum_{u=1}^{n_0}(\lambda_u^A - \overline{\Lambda^A})^2}\sqrt{\sum_{v=1}^{n_0}(\lambda_v^B - \overline{\Lambda^B})^2}} \quad (14)$$

其中权值 $w_{uv} > 0$，S 为向量 Λ^A 的协方差矩阵，其大小为 $n_0 \times n_0$，$\overline{\Lambda^A}$ 和 $\overline{\Lambda^B}$ 分别为特征向量 Λ^A 和 Λ^B 的均值。

5 实验结果与分析

本文在实验环境 Intel Core 2.0G、2.0G 内存、windowsXP, Matlab7.01 下进行。首先进行改进 PCNN 二值序列图像 NMI 特征矢量提取及抗几何畸变性实验，然后再将 NMI 不变矢量特征应用在图像检索中。实验中采用国际上通用的 Corel 图像库，其中包括交通工具、动物、建筑物、自然景物、花卉及不同纹理等类型的图像 3000 多幅，从中选取 6 大类且每类包含 6 小类共 36 种分组集合。实验中不同图像在同一 PCNN 模型下进行，参数设定是 $\beta = 0.1, V_L = 0.05, \alpha_T = 0.072, V_T = 400, n_0 = 35$，内部连接矩阵 $W = [0.707\ 1\ 0.707;1\ 1\ 1;0.707\ 1\ 0.707]$。

(1) 任选多幅图像在旋转、缩放及平移的情况下运用 PCNN 模型提取 NMI 特征矢量进行抗几何畸变性实验，文中仅选列图 1(a) bridge 和图 2(a) horse 两幅图像的实验结果。图 2(b) ~ (f) 分别为原图像生成的二值序列图像 NMI 特征矢量序列、旋转 45°、放大 1.2 倍、缩小 0.8 倍及平移后提取的 NMI 特征矢量序列，表 1 分别为两图像在采用传统 PCNN 和改进 PCNN 时提取 NMI 的综合相似性度量及处理时间的数据结果。

表1 采用两种 PCNN 模型提取原图像及几何畸变图像 NMI 特征矢量序列综合相似性度量和处理时间的对比数据

方法	图像	旋转		放大		缩小		平移	未畸变的原图	平均处理时间(s)
		20°	45°	1.2倍	1.4倍	0.6倍	0.8倍			
传统 PCNN	bridge	0.04377	0.11320	0.07750	0.06920	0.06240	0.07250	0.00190	0.00000	19.1
	horse	0.00366	0.01391	0.00662	0.04783	0.00890	0.00370	0.00040	0.00000	
改进 PCNN	bridge	0.03150	0.08050	0.05060	0.04610	0.04840	0.05830	0.00140	0.00000	3.80
	horse	0.00271	0.00983	0.00520	0.00200	0.00660	0.00190	0.00020	0.00000	

由图 1 可见，利用改进 PCNN 对原始图像处理可获得图像间既有差别又具相关性的二值系列图像，且该序列能代表和体现原图像的特征信息；由图 2 可知当图像发生旋转、缩放及平移等几何畸变时提取的二值序列图像具有相似的 NMI 特

征矢量形状,虽然对不同的畸变,NMI 特征矢量值大小有所不同,但其特征矢量的曲线形状及走势是不变的;另外,从表1实验结果可以得出当分别采用传统与改进 PCNN 处理来提取图像的 NMI 特征,在一定误差范围内其综合相似性度量 C 值和特征提取时间后者均小于前者,主要是改进 PCNN 由于去除了传统模型中相近神经元间一些繁冗复杂的耦合以及改进了反复衰变的动态阈值,使得二值序列图像中许多重要特征信息能较多较快稳定的凸现出来,同时实验结果充分说明了由此提取的 NMI 特征矢量具有良好的抗几何畸变不变性。

(2)为验证本文所提算法的有效性并检验图像检索结果的一般性,从图像库中任意取出 10 类,每类图像中选取 10 幅图像作为示例图像,共组成 100 次查询,取这 100 次检索结果精确度(precision)和检索率(recall)[22]的平均值作为算法相似检索评价准则来检验其平均检索结果,并与文献[17]和[18]的检索算法进行了比较,图 3 给出了本文算法同其它两种算法的 P – R 对比曲线。对不同图像检索算法而言,在相同检索条件下检索的精确度越高表明算法的检索效果越好。

为进一步检验所提算法的性能,我们采用平均归一化调整后的检索秩(Average Normalized Modified Retrieval Rank,ANMRR)[17]和图像检索的计算时间来进行评测。设查询图像 q_i 的相似图像个数为 $ng(q_i)$,$i=1,2,\cdots,Q$,GTM 是在所有查询图像中最大相似图像个数,即 $GTM = \max\{ng(q_i)\}$,$K = \min\{4ng(q_i), 2GTM\}$ 为检索结果的截断值,设与查询图像相似的图像在检索结果序列中所处位置为:

$$rank(l) = \begin{cases} l, & l \leq K \\ l+1, & l > K \end{cases} \tag{15}$$

则 ANMRR 的定义如下:

$$ANMRR = \frac{1}{Q}\sum_{i=1}^{Q}\frac{\sum_{l=1}^{ng(q_i)}\frac{rank(l)}{ng(q_i)} - 0.5 - 0.5ng(q_i)}{K + 0.5 - 0.5ng(q_i)} \tag{16}$$

ANMRR 值越低,查准性能越好,说明更多正确的结果排在前面。表 2 给出了三种不同算法 ANMRR 和单次图像平均检索时间的评测结果。

由实验图表可以看出本文算法在检索计算时间远小于文献[17]而与文献[18]算法相当的情况下,利用符合人类视觉处理 PCNN 的相似时空捕获特性来二值化图像,所得二值序列图像充分保留了原图像尽可能多目标信息,而由此提取的 NMI 矢量序列可充分表达和反映不同图像的特征,取得了较好检索效果;而文献[17]中只把原图像按灰度等间隔进行二值量化,没有考虑序列图像中邻域像素的空间相似性以及不同图像间的相关性,二值后序列图像中只是一些离散像素点,没有形成有效目标区域,导致不能较好提取 NMI 特征值并用于检索。文献

[18]从位平面的角度对图像进行了二值处理,并通过提取高四位平面欧拉矢量的不同组合进行图像检索,这种方法虽然简单,但没有充分挖掘出原图像较多重要信息,信息损失量较大,影响到图像检索效果。总之,由图3可以看出在相同检索率的情况下,本文所提算法检索结果精确度要明显高于其他两种算法的精确度(例如在检索率为0.3时,文献[17]的精确度为0.79,文献[18]的精确度为0.80,而本文精确度为0.87),并且随着检索率的增加,三种算法的检索精确度都有所下降,但本文所提算法一直保持较高的检索精确度,同时由表2的ANMRR实验数据表明所提算法对图像正确检索结果都要高于其他的两种算法,体现了它的查准性能较好。

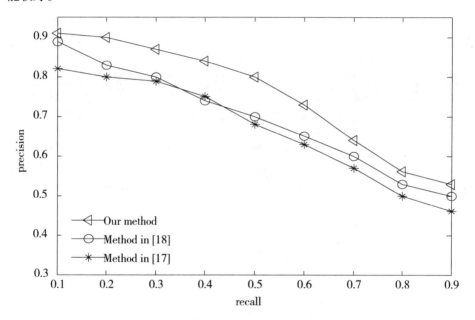

图3　3种不同算法的检索率与精确度对比曲线

表2　3种不同算法的ANMRR及图像平均检索时间

算法	ANMRR	平均检索时间(s)
Method in [17]	0.3056	83.92
Method in [18]	0.2253	3.170
Our method	0.1818	3.800

9

(3)为验证综合相似性度量的有效性,实验中对不同的相似性度量方法进行了对比实验。图4给出了利用改进PCNN二值处理提取NMI特征后分别应用Pearson积矩相关法、马氏距离和本文综合相似性度量进行相似性度量实验中一次关于"桥与建筑"的检索结果,其中最左边的一幅为示例图像兼检索结果,随后为检索结果图像队列,图4(a)中其相关值从左到右、从上到下逐渐减小,图4(b)、(c)马氏距离值和综合相似性度量值从左到右、从上到下逐渐增大,但其相似度都是由大到小排列。由实验结果可以看出,由于综合相似性度量考虑了数据特征矢量间的非线性、非正态分布、相关及数据局部等特性,有效地克服了相关法和马氏距离的一些不足,其检索效果要优于Pearson积矩相关法(出现明显的两个错检)和马氏距离(出现明显的一个错检)相似性度量,检索结果更符合人类的视觉特性。

(a) Pearson积矩相关法的实验结果

(b) 马氏距离的实验结果

(c) 综合相似性度量的实验结果

图4 3种不同的相似性度量方法的实验结果

6 结论

本文提出了一种基于脉冲耦合神经网络图像归一化转动惯量特征提取及检索算法。在利用改进PCNN模型相似神经元同步时空特性及指数衰降机制的基础上,将图像分解成图像间有一定相关性的二值系列图像,且通过提取序列中各二值图像的一维 NMI 矢量信号对原图像特征进行描述并应用在图像检索中,同时,引入了马氏距离结合相关性的综合相似性度量方法。实验结果表明,本文算法经改进 PCNN 处理的 NMI 矢量特征充分考虑了图像目标及序列间相关性,体现了良好抗几何畸变不变及表述唯一的特性,并运用综合相似性度量方法,达到了较好的图像检索效果。

参考文献

[1] Krishnapuram, R. Medasani, S. Sung – Hwan Jung, et al. Content – based image retrieval based on a fuzzy approach[J]. IEEE Transactions on Knowledge and Data Engineering, 2004, 16(10):1185 – 1199.

[2] Dr. N. Krishnan, M. Sheerin Banu, C. Callins Christiyana. Content Based Image Retrieval using Dominant Color Identification Based on Foreground Objects[A]. 2007 International Conference on Computational Intelligence and Multimedia Applications[C], Sivakasi, India, 2007, 3(13 – 15):190 – 194.

[3] Kohonen, Oili; Hauta – Kasari, Markku. Distance measures in the training phase of self – organizing map for color histogram generation in spectral image retrieval[J]. Journal of Imaging Science and Technology, 2008, 52(2):201 – 211.

[4] Aleksandra Mojsilovic, Jianying Hu, Emina Soljanin. Extraction of Perceptually Important Colors and Similarity Measurement for Image Matching, Retrieval, and Analysis[J]. IEEE TRANSACTIONS ON IMAGE PROCESSING, 2002, 11(11):1238 – 1248

[5] Gagaudakis, George; Rosin, Paul L. Shape measures for image retrieval[J]. Pattern Recognition Letter, 2003, 24(15):2711 – 2721.

[6] Arijit Bishnu, Bhargab B, Bhattacharya, et al. Euler Vector for Search and Retrieval of Gray – Tone Images[J]. IEEE TRANSACTIONS ON SYSTEMS, MAN, AND CYBERNETICS—PART B:CYBERNETICS, 2005, 35(4):801 – 811

[7] Wong Wai – Tak; Shih Frank Y. ; Su T. E. – Feng. Shape – based image retrieval using two – level similarity measures[J]. International Journal of Pattern Recognition and Artificial Intelligence, 2007, 21(6):995 – 1015.

[8] Naif Alajlan, Mohamed S. Kamel, George H. Freeman. Geometry – Based Image Retrieval in Binary Image Databases[J]. IEEE TRANSACTIONS ON PATTERN ANALYSIS AND MACHINE INTELLIGENCE, 2008. 30(6):1003 – 1013.

[9] Fei Li, Qionghai Dai, Wenli Xu, et al. Multilabel Neighborhood Propagation for Region – Based Image Retrieval[J]. IEEE TRANSACTIONS ON MULTIMEDIA, 2008, 10(8):1592 – 1604.

[10] Chun Young Deok, Kim Nam Chul, Jang Lck Hoon. Content – based image retrieval using

multiresolution color and texture features[J]. IEEE Transactions on Multimedia,2008,10(6): 1073 - 1084.

[11]M. Pi,H. Li. Fractal indexing with the joint statistical properties and its application in texture image retrieval[J]. IET Image Process. ,2008,2(4):218 - 230.

[12]R Eckhorn,H J Reitboeck,M Arndtetal. Feature linking via synchronization among distributed assemblies:simulation of results from cat cortex [J]. Neural Computation,1990,2(3): 293 - 307.

[13]J L Johnson,M L Padgett. PCNN Models and Applications [J]. IEEE,Trans,Neural Networks,1999,10(3):480 - 498.

[14]Liu Qing,Ma Yi - de. Qian Zhi - Bai. Automated image segmentation using improved PCNN model based on cross - entropy [J]. Journal of Image and Graphics,2005,10(5):579 - 584.

刘勍,马义德,钱志柏.一种基于交叉熵的改进型PCNN图像自动分割新方法[J].中国图象图形学报.2005,10(5):579 - 584.

[15]Liu Qing,Ma Yi - de. A New algorithm for Noise Reducing of Image Based on PCNN Time Matrix[J]. Journal of Electronics & Information Technology,2008,30(8):1869 - 1873.

刘勍,马义德.一种基于PCNN赋时矩阵的图像去噪新算法[J].电子与信息学报,2008,30(8):1869 - 1873.

[16]Liu Qing,Ma Yi - de,Zhang Shao - gang,et al. Image Target Recognition Using Pulse Coupled Neural Networks Time Matrix[A]. Proceedings of the 26th Chinese Control Conference [C]. Zhangjiajie,Hunan,China. 2007,7:96 - 99.

[17]ZHANG Wen - yin,TANG Jian - guo,ZENG Zhen - bi. Image Retrieval Method Based on NMI Invariable Feature[J]. Computer Application,2003,23(9):55 - 57.

张问银,唐建国,曾振柄.基于NMI不变特征的图像检索方法[J].计算机应用.2003,23 (9):55 - 57.

[18]Arijit Bishnu,Bhargab B. Bhattacharya,Sr. Stacked Euler Vector(SERVE):A Gray - Tone Image Feature Based on Bit - Plane Augmentation[J]. IEEE Trans,Pattern Analysis and Machine Intelligence,2007,29(2):350 - 355.

[19]Zhao Shan,Cui Jiang - tao,Zhou Li - hua. Image Retrieval Based on Bit - plane Distribution Entropy[J]. Journal of Electronics & Information Technology,2007,29(4):795 - 799.

赵珊,崔江涛,周利华.基于位平面分布熵的图像检索算法[J].电子与信息学报,2007, 29(4):795 - 799.

[20]Weng Muyun,He Mingyi. Image Feature Detection and Matching Based on SUSAN Method[A]. Proceedings of the First International Conference on Innovative Computing,Information and Control [C]. Washington,DC,USA,2006,322 - 325.

[21]Anderson T W. An Introduction to Multivariate Statistical Analysiys. New York:Wiley, 2003,chapter3.

[22] Smeulders A W, Santini S, and Worring M, et al.. Content based image retrieval at the end of the early years[J]. IEEE Trans. on Pattern Analysis and Machine Intelligence, 2000, 22(12):1349 – 1380.

注:本文曾发表在 2010 年 7 月《自动化学报》第 7 期上。

结合灰度熵变换 PCNN 小目标图像检测新方法

刘 勍[*]

为了自动地进行小目标图像分割检测,从含单一弱小目标图像的特征出发,提出了一种结合灰度熵变换的脉冲耦合神经网络(PCNN)小目标图像分割检测新方法。该方法在对含随机噪声和复杂背景图像进行非线性灰度熵变换滤波的基础上,考虑灰度熵值灰度图在满足先验概率目标背景比条件下,选择包含单一小目标局部窗口作为处理图像区域,并在局部最小交叉熵判据下,进行改进型 PCNN 迭代分割检测处理。实验表明,该方法不仅能可靠地检测出复杂背景及随机噪声干扰下的弱小目标,而且在 PCNN 运行处理过程中,可自动地完成最佳分割检测,具有较强的适用性和良好的目标检测能力。

引言

在卫星遥感、航空航天、医学生物工程等领域,弱小目标图像检测及识别是应用前景比较广泛的一种前沿技术。而弱小目标图像分割检测又是数字图像处理中一个极具挑战性的急需解决难题,特别在红外图像小目标分割检测中由于小目标距离远、占像素数少、运动速度低,而且缺乏形状、尺寸、纹理等信息,同时,由于存在着各类干扰,造成图像的信噪比低,对目标进行检测与跟踪十分不易,其中最为困难的就是准确自动地进行目标的分割[1]。目前,在已有的图像小目标分割检测方法中,归纳起来可分为时域处理与空域处理两大类。在时域处理中,主要依据目标在时间上灰度与运动的连续性进行分割,在图像序列中相邻的几帧进行时域分析,其处理算法是基于不同时间采集到的多帧图像,且序列必须配准,不适用于单帧图像目标检测,主要有基于序列图像检测法[2~4]、时域滤波法[5~6],基于小波变换动态背景检测法等[7];空域处理主要利用目标在单帧图像的灰度分布特

---

[*] 作者简介:刘勍(1970—),男,甘肃天水人,天水师范学院教授、博士,主要从事图像信号处理及人工神经网络研究。

性,对图像空域特性进行分析,获取目标和背景的差异,再通过后续相关处理完成最终目标检测,典型的方法有高通滤波法[8]、局部熵法[9]、形态学滤波法[10]、空间滤波法[11]和局部邻域统计[1,12]等方法。上述方法在目标检测中有一定的实用性,但对部分缓变背景和含噪图像的小目标检测性能有限,而且在复杂背景下,不能有效抑制噪声干扰,不能自动地分割检测目标,实时处理性能和适应性差。

脉冲耦合神经网络(pulse coupled neural networks,PCNN)是一种不同于传统人工神经网络的新型神经网络。因为 PCNN 有着生物学的背景,它是依据猫、猴等动物大脑视觉皮层上的同步脉冲发放现象提出的[13],已被广泛地应用于图像平滑、分割、边缘检测及目标识别等图像处理领域[14~17],并显示了其优越性。本文首次将 PCNN 应用于图像弱小目标的分割检测中,提出了基于灰度熵变换 PCNN 图像小目标分割检测方法。首先,对小目标图像进行灰度熵变换,然后利用图像局部统计特性,选取含小目标局部图像区域,运用 PCNN 迭代处理并采用局部最小交叉熵准则来自动快速地判决分割检测图像,且当满足最小交叉熵时,完成最终目标的自动分割,最后通过实验验证了所提算法的有效性。

1 图像灰度熵变换

熵是信息论中事件出现概率不确定性的量度,它能有效反映事件包含的信息,而图像像素灰度反映了能量的空域分布,把它与信息论中熵的概念结合并推广到图像处理领域,可得出图像灰度熵的概念[18],作为能量空域分布状态不确定性的量度。设 $f(i,j)$ 为图像中点 (i,j) 处的灰度,对于一幅 $M \times N$ 大小的图像,图像灰度熵 H_f 定义为

$$H_f = - \sum_{i=1}^{M} \sum_{j=1}^{N} p_{ij} \log p_{ij} \qquad (1)$$

其中

$$p_{ij} = f(i,j) / \sum_{i=1}^{M} \sum_{j=1}^{N} f(i,j) \qquad (2)$$

式中 p_{ij} 为图像的灰度分布。如果 $M \times N$ 是图像的局部窗口,则称 H_f 为图像的局部熵。图像灰度熵表征了图像能量分布的宏观统计特征,反映了窗口内像素的差异程度。灰度熵大的地方,表明图像灰度相对比较均匀,灰度熵小的地方,图像灰度混乱程度较大;另外,灰度熵具有抗几何失真和旋转不变的特性。由(1)可知,图像灰度熵运算(即熵变换)是一种非线性空域滤波,在红外图像或天文小目标图像中由于图像背景是均匀或缓变的且图像本底噪声近似为随机噪声,则其灰度熵近似相等,而图像中有目标出现的地方,图像灰度有突变,其灰度熵也会发生突变。

对于一幅 $M \times N$ 大小的图像,要进行图像灰度熵变换,首先选定邻域为 $k \times l$ 尺寸的窗口,然后逐个像素计算以当前像素为中心的邻域窗口范围内图像灰度熵值,可得灰度熵值分布图,最后将图像灰度熵值分布图按照比例分布范围,比例映射到 $[G_1,G_2]$ 图像灰度空间,并进行反色处理,得到局部灰度熵值的灰度图显示。

2 PCNN 模型及其分割检测算法

2.1 PCNN 基本模型及其改进

PCNN 也称为第三代人工神经网络,它是在生物视觉皮层模型的启发下产生由若干个神经元互连而构成的反馈型网络。构成 PCNN 的每一神经元由接收、调制和脉冲产生 3 部分组成,其离散数学方程描述如下[15]:

$$F_{ij}[n] = e^{-\alpha_F} F_{ij}[n-1] + V_F \sum_{kl} M_{ijkl} Y[n-1] + I_{ij} \quad (3)$$

$$L_{ij}[n] = e^{-\alpha_L} L_{ij}[n-1] + V_L \sum_{kl} W_{ijkl} Y_{kl}[n-1] \quad (4)$$

$$U_{ij}[n] = F_{ij}[n](1 + \beta L_{ij}[n]) \quad (5)$$

$$T_{ij}[n] = e^{-\alpha_T} T_{ij}[n-1] + V_T Y_{ij}[n] \quad (6)$$

$$Y_{ij}[n] = \begin{cases} 1, (U_{ij}[n] > T_{ij}[n]) \\ 0, (U_{ij}[n] \leq T_{ij}[n]) \end{cases} \quad (7)$$

其中 ij 下标为神经元的标号,n 为迭代次数,I_{ij}、$F_{ij}[n]$、$L_{ij}[n]$、$U_{ij}[n]$、$T_{ij}[n]$ 分别为神经元的外部刺激(图像像素构成的矩阵中第 ij 个像素的灰度值)、第 ij 个神经元的第 n 次反馈输入、连接输入、内部活动项和动态阈值,M 和 W 为链接权矩阵,V_F、V_L、V_T 分别为 $F_{ij}[n]$、$L_{ij}[n]$ 及 $T_{ij}[n]$ 的幅度常数,α_F、α_L、α_T 为相应的衰减系数,β 为链接系数,$Y_{ij}[n]$ 是 PCNN 的二值输出。

PCNN 进行图像处理时,首先将一个 2 维 PCNN 网络的 $M \times N$ 个神经元分别与 2 维输入图像的 $M \times N$ 个像素相对应,若 I_{ij} 大于阈值,则这时神经元输出为 1,称为激活,被激活的神经元则通过与之相邻神经元的连接作用激励邻接神经元,若邻接神经元的内部活动项大于阈值,则被捕获激活。显然,如果邻接神经元与前一次迭代激活的神经元所对应的像素具有相似强度,则邻接神经元容易被捕获激活,反之则不能被捕获激活。因此,利用某一神经元的激活会触发它邻域内相似神经元的集体激活,其对应于图像中有相似性质一小区域特性,可进行图像的分割。

由于传统 PCNN 模型主要体现在非线性调制耦合和阈值指数衰变方面,而非线性调制耦合是 PCNN 的核心。传统 PCNN 模型的阈值虽然是指数衰降的,但又是反复变化的,即经过长时间(或短时间)的衰降后必然会有由于神经元激活造成的突然上升,之后又开始衰降,然后又突然上升……,很显然,这种阈值变化规律不能符合人眼对亮度响应的非线性指数要求,并且通过这种阈值规律处理后的图像(或其他信号)中大量信息蕴含在神经元的激活周期(频率)或者激活相位中,而输出的二值图像却并不包含全部的可用信息。为克服上述缺点,且从小目标图像特性出发,本文对阈值衰减函数进行了改进,即把随时间反复衰变的指数函数改进为随时间单调递减的指数函数并对模型进行了简化,其改进简化模型表示如下:

$$F_{ij}[n] = I_{ij} \tag{8}$$

$$L_{ij}[n] = V_L \sum_{kl} W_{ijkl} Y_{kl}[n-1] \tag{9}$$

$$T_{ij}[n] = T_0 e^{-\alpha_T(n-1)} \tag{10}$$

$$Y_{ij}[n] = \begin{cases} F_{ij}, & (U_{ij}[n] > T_{ij}[n]) \\ Y_{ij}[n-1], & (U_{ij}[n] \leq T_{ij}[n]) \end{cases} \tag{11}$$

其中 $U_{ij}[n]$ 的表达式与式(5)相同,改进型 PCNN 用于图像小目标分割检测时,有以下关键概念:(1)神经元的外部输入是与之相关联像素的灰度值,即 $F_{ij}[n] = I_{ij}$;(2)所有神经元结构相等,且各个神经元的参数一致;(3)每个神经元接受与之距离 R 以内的神经元链接输入,其内部链接矩阵 W 是一个 3×3 的方阵,且每一个元素的值为中心像素到周围每个像素的欧几里德距离的倒数;(4)每个神经元只能激活一次。

2.2 最小交叉熵阈值分割

交叉熵是用于度量2个概率分布之间信息量的差异[19],是一个下凸函数,而最小交叉熵准则应用于图像分割中,一般是搜索使分割前后图像信息量差异最小的阈值。设有两个概率分布 $P = \{p_1, p_2\cdots, p_N\}$ 和 $Q = \{q_1, q_2\cdots, q_N\}$,用交叉熵度量它们之间的信息量差异,定义为

$$C(P:Q) = \sum_{i=1}^{N} p_i \ln \frac{p_i}{q_i} \tag{12}$$

将分割前后图像中像素特征矢量的概率分布分别用 P, Q 表征,求最优阈值使原始图和分割图之间的信息量差异最小,就可得到最小交叉熵阈值化分割算法。其对称交叉熵定义如下

$$C(P,Q;t) = \sum_{f=0}^{t} [fh(f)\ln\frac{f}{\mu_o(t)} + \mu_o(t)h(f)\ln\frac{\mu_o(t)}{f}]$$

$$+ \sum_{f=t+1}^{G} [fh(f)\ln\frac{f}{\mu_b(t)} + \mu_b(t)h(f)\ln\frac{\mu_b(t)}{f}] \quad (13)$$

其中

$$\mu_o(t) = \sum_{f=0}^{t} fh(f) / \sum_{f=0}^{t} h(f), \quad \mu_b = \sum_{f=t+1}^{G} fh(f) / \sum_{f=t+1}^{G} h(f)$$

式(13)中 f 是图像灰度值，$h(f)$ 是图像的灰度统计直方图，G 是灰度上界，t 是假定的阈值，$\mu_o(t)$ 和 $\mu_b(t)$ 是类内均值。在计算中，用 G 对式(13)进行归一化处理，由于它是在假定一个阈值情况下的原图像和分割结果图像之间的信息量差异度量结果，所以可以在图像灰度范围内搜索 t 值，而使式(13)最小的 t 即为最佳分割阈值 t_0。

$$t_0 = \mathrm{argmin}(C(P,Q:t)) \quad (14)$$

如果对图像中某一局部窗口区域采取交叉熵分割，则其为局部交叉熵。

2.3 小目标图像自动分割检测算法

当经过灰度熵变换后含单一目标灰度熵值灰度图的目标区较小，则其直方图不能形成任何波峰，其重要信息只存在高亮度的小目标区域，如果对一整幅图像逐像素进行扫描分割，常规的方法无法满足分割要求，必然造成算法的运行时间加长，既不利于实时处理，也会容易引起错误分割。为此本文在灰度熵值灰度图的基础上，利用小目标的统计概率特性，考虑目标像素和背景像素在图像中出现的先验概率比，即 $p_o(o|f)/p_b(b|f)$（其中 f 为图像上某点的观测值，o、b 分别表示该像素点为目标和背景）。当 $(p_o(o|f)/p_b(b|f)) > \lambda$ 时，该像素点为目标点；当 $(p_o(o|f)/p_b(b|f)) < \lambda$ 时，该像素点为背景点（λ 设为决策门限）。据此以灰度熵灰度图中最大像素为中心选取一满足大于 λ 条件的局部处理区域（一般为方形窗口图像区域），并对该区域进行 PCNN 迭代处理，同时运用局部最小交叉熵准则进行判决。具体分割检测算法描述如下：

(1) 输入原始图像；

(2) 进行图像灰度熵变换，生成灰度熵值分布图；

(3) 对灰度熵值分布图进行比例映射，并反色处理，生成灰度熵值灰度图；

(4) 应用统计概率目标背景比条件，当满足大于决策门限 λ 时，选取待处理含小目标局部图像区域，未选中的其他图像区域置 0 处理；

(5) 对局部图像区域进行改进 PCNN 迭代处理，并在迭代过程中执行局部交叉熵判据，当满足局部交叉熵最小时，自动得出最佳分割阈值 t_0，并生成迭代次数，产生局部分割图像。

(6) 输出总的分割图像。

3 实验结果与分析

利用 MATLAB 仿真环境,我们对大量含单一小目标图像进行实验处理,验证了所提算法的有效性。在此选取两幅具有代表性的序列单帧图像,其中一幅为 238×238 缓变背景叠加随机噪声的红外弱小目标图像,如图 1(a)所示,另一幅为 143×107 的具有复杂背景的含团块小目标光学图像,如图 2(a)所示。实验中选用的灰度熵变换窗口为 5×5,其他参数是 $\beta = 0.1, V_L = 0.05, T_0 = 255, \alpha_T = 0.09, \lambda_{红} = 0.001, \lambda_{光} = 0.01$。

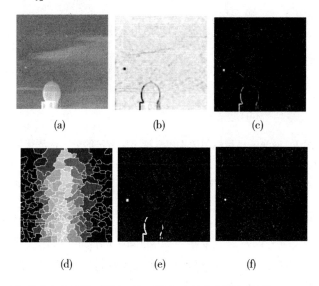

图1 对红外小目标图像采用三种不同方法的分割检测结果.
(a)原始图像;(b)图像灰度熵值分布图;(c)灰度熵值灰度图;
(d)分水岭分割;(e)均值方差阈值法分割;(f)本文方法分割

(1)对图 1(a)红外图像进行本文分割检测算法处理。选取满足大于 $\lambda_{红} = 0.001$ 的以最大像素为中心 17×17 灰度熵值灰度局部区域图像进行 PCNN 迭代处理。图 1(b)为图像灰度熵值分布图,(c)为映射到[0,255]的灰度熵值灰度图,(d)为基于分水岭分割图,(e)为均值方差阈值法分割图,而(f)为本文方法分割的局部最小交叉熵 $C_{min} = 0.0914$,迭代次数 $N_{min} = 15$ 时的二值图像;

(2)对图 2(a)光学图像进行本文分割检测算法处理。选取满足大于 $\lambda_{光} = 0.01$ 的以最大像素为中心 25×25 灰度熵值灰度局部区域图像进行 PCNN 迭代处理。图 2(b)为图像灰度熵值分布图,(c)为映射到[0,255]的灰度熵值灰度图,(d)为基于分水岭分割图,(e)为均值方差阈值法分割图,(f)为采用本文方法分

割的局部最小交叉熵 $C_{min}=0.0909$,迭代次数 $N_{min}=22$ 时的二值图像;

(3)表1以图1(a)原始红外图像为实验对象,列出了在给定决策门限 $\lambda_{红}=0.001$ 时,当选择含小目标的不同局部区域图像(其中 5×5 和 7×7 局部处理区域不满足 $\lambda_{红}=0.001$ 条件)进行 PCNN 处理的最佳分割检测对比数据。

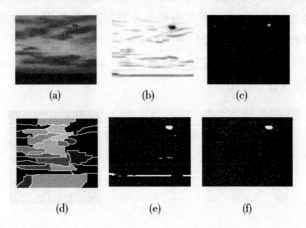

图2 对光学小目标图像采用三种不同方法的分割检测结果.
(a)原始图像;(b)图像灰度熵值分布图;(c)灰度熵值灰度图;
(d)分水岭分割;(e)均值方差阈值法分割;(f)本文方法分割

从图1、图2中的(b)、(c)图可以看出,对原始图像经过图像灰度熵变换,有效地去除了均匀、缓变以及部分复杂背景,滤除了图像本底的随机噪声(图像中如含有其他诸如脉冲噪声等,可在灰度熵变换前先进行中值滤波等处理),虽然在灰度熵值灰度图中,小目标的边缘范围有所扩散,但不影响含小目标区域以较高灰度值体现出来的特征,为后续处理奠定了基础。两图中(d)的基于分水岭分割方法产生了完全错误的分割结果,不能分割出小目标来,两图中(e)均值方差阈值法分割虽然能分割出小目标来,但是图像中明显地产生了部分错误分割,分割效果不理想,而图1、图2中(f)本文方法分割的二值图像中均能分割出小目标来,且分割效果理想;从表1可以看出,当选择满足大于决策门限 λ 的含小目标局部不同处理区域时,由局部最小交叉熵判决处理区域的大小与迭代次数及分割阈值无关,而不满足 λ 条件决定的区域由于没有完全包含全部的目标区域,产生了不同的变化(迭代次数和分割阈值),但从算法的处理速度和实际处理的要求出发,不能把局部处理区域选取太大;此外,由于本文算法执行了 PCNN 迭代处理,且运用了局部最小交叉熵准则判据,加强了自动分割检测的能力。

表1 当选取不同区域大小时红外小目标图像的最佳分割检测数据

局部处理区域大小	5×5	7×7	9×9	13×13	17×17	21×21	25×25	29×29	33×33	37×37	41×41
最小交叉熵值 C_{min}	0.1589	0.2248	0.2033	0.1355	0.0914	0.0678	0.0543	0.0464	0.0408	0.0376	0.0352
PCNN 代次数 N_{min}	8	16	15	15	15	15	15	15	15	15	15
最佳分割阈值 t_0	132.6	64.55	70.63	70.63	70.63	70.63	70.63	70.63	70.63	70.63	70.63

4 结束语

本文提出了一种结合灰度熵脉冲耦合神经网络的小目标图像分割检测新方法。该方法首先对有一定背景以及本底为随机噪声的图像进行一种非线性空域滤波—灰度熵变换,然后对灰度熵值灰度图在满足先验概率目标背景比条件下,选择分离出含有单一小目标的局部图像处理区域,在局部最小交叉熵准则的判据下,进行改进 PCNN 迭代分割检测处理。实验结果表明该方法不仅能可靠地检测出缓变复杂背景以及随机噪声干扰下的弱小目标,而且算法在运行过程中不需人工干预,可自动快速地完成最佳分割检测,具有较强的适用性和良好的应用前景。而小目标图像检测一直是具有挑战性的研究难题,如何能较好的克服运用灰度熵变换引起的小目标边缘的轻微扩散、利用小目标的特性,进一步更快更好地完成决策门限的确定以及对多小目标图像自动快速分割检测处理等将是进一步要研究的内容。

参考文献

[1] XU Bin,ZHENG Lian,WANG Ke‐yong,et al. Dim TargetsDetection Based on LocalGray Probability Analysis[J]. LASER & INFRARED(许彬,郑链,王克勇,等. 基于局域灰度概率分布的小目标检测方法[J]. 激光与红外),2005,35(3):187‐189.

[2] Courtney I Hilliard. Selection of a clutter rejection algorithm for real‐time target detection from an air‐borne platform[C]. In SPIE Proceedings:signal and data Proeessing of small targets

[A]. Orlando, FL. USA. 2000, 4048. 74 – 84.

[3] WU Wei, PENG Jia – xiong, LIU Quan. Re search o n Segmenting Small Target in the Infrared Image Sequences[J]. ACTA ELECTRONICA SINICA(吴巍, 彭嘉雄, 刘泉. 对红外序列图像中小目标分割的研究[J]. 电子学报), 2004, 32(7): 1116 – 1119.

[4] XU Jian – feng, WU Yi – quan, ZHOU Jian – jian. Small Target Detection Based on Temporal Predictions of Background in Infrared Image Sequences[J]. Journal of Image and Graphic(徐剑峰, 吴一全, 周建江. 基于时域背景预测检测红外图像序列中的小目标[J]. 中国图象图形学报), 2007, 12(9): 1598 – 1603.

[5] Alexander Tartakovsky. Effective adaptive spatial – temporal technique for clutter rejection in IRST[C]. In SPIE Proceedings: Signal and Data Proceeding of Small Targets[A]. Orlando, FL. USA. 2000, 4048. 85 – 95.

[6] Wei Zhang Mingyu Cong Liping Wang. ALGORITHMS FOR OPTICAL WEAK SMALL TARGETS DETECTION AND TRACKING: REVIEW[C]. IEEE Int. Conf. Neural Networks & Signal Processing [A]. Nanjing, China. 2003, 643 – 647.

[7] Qing – bo JI, Chi FENG, Meng WANG. A DETECTION ALGORITHM FOR THE SMALL MOVING TARGET IN INFRARED IMAGE SEQUENCES WITH THE DYNAMIC BACKGROUND[C]. Proceedings of the 2007 International Conference on Wavelet Analysis and Pattern Recognition[A]. Beijing, China, 2007, 1717 – 1722.

[8] L. Yang, J. Yang, K. Yang. Adaptive detection for infrared small target under sea – sky complex background[J]. ELECTRONICS LETTERS, 2004, 40(17): 1083 – 108.

[9] ZHOU Bing, WANG Yong – zhong, SUN Li – hui, te al. Study on Local Entropy Used in Small Target Detectio[J]. ACTA PHOTONICA SINICA(周冰, 王永仲, 孙立辉, 等. 图像局部熵用于小目标检测研究[J]. 光子学报), 2008, 37, (2): 381 – 387.

[10] WEN Pei – zh, i SHI Ze – lin, YU Hai – bin. A detection method for IR point target on sea background based on morphology[J]. Opto – ElectronicEngineering. 2003, 30(6): 55 – 58.

[11] LUO Jun – Hui, JI Hong – Bing, LIU Jin. ALGORITHM OF IR SMALL TARGETS DETECTION BASED ON SPATIAL FILTER AND ITS APPLICATION[J]. J. Infrared Millim. Waves(罗军辉, 姬红兵, 刘靳. 一种基于空间滤波的红外小目标检测算法及其应用[J]红外与毫米波学报), 2007, 26(3): 209 – 212.

[12] WANG Xun, BI Duyan. Dim Targets Detection Based on Local Character Information Measurement[J]. Computer Engineering(王勋, 毕笃彦. 一种新的基于局部特征统计的小目标检测方法[J]. 计算机工程), 2007, 33(12): 19 – 3.

[13] R Eckhorn, H J Reitboeck, M Arndtetal. Feature linking via synchronization among distributed assemblies: simulation of results from cat cortex [J]. Neural Computation, 1990, 2(3): 293 – 307.

[14] J L Johnson, M L Padgett. PCNN Models and Applications [J]. IEEE, Trans, Neural Net-

works,1999,10(3):480~498.

[15] Liu Qing, Ma Yi-de. A New algorithm for Noise Reducing of Image Based on PCNN Time Matrix[J]. Journal of Electronics & Information Technology(刘勍,马义德.一种基于PCNN赋时矩阵的图像去噪新算法[J].电子与信息学报),2008,30(8):1869-1873.

[16] MA Yi-de, LIU Qing, QIAN Zhi-bai. Automated Image Segmentation Using Improved PCNN Model Based on Cross-entropy[C]. Proceedings of 2004 International Symposium on Intelligent Multimedia, Video and Speech Processing [A]. Hong Kong. 2004-10:743~746.

[17] LIU Qing, MA Yi-de, ZHANG Shao-gang, et al. Image Target Recognition Using Pulse Coupled Neural Networks Time Matrix[C]. Proceedings of the 26th Chinese Control Conference [A]. Zhangjiajie, Hunan, China. 2007,7:96~99.

[18] ZHANG Yong-liang, WANG Yang, LU Huan-zhang. Block objects detection based on entropy of brightness[J]. Systems Engineering and Electronics(张永亮,汪洋卢,焕章.基于图像灰度熵的团块目标检测方法[J].系统工程与电子技术),2008,30(2):201-204.

[19] Xue Jinghao, Zhang Yujin, Lin Xinggang. Image thresholding based on maximum between-class posterior cross entropy [J]. Journal of Image and Grophics,1999,4(2):110~114.(薛景浩,章毓晋,林行刚.基于最大类间后验交叉熵的阈值化分割算法[J].中国图象图形学报),1999,4(2):110-114.

注:本文曾发表在2009年《北京理工大学学报》第12期上

一种基于超模糊熵 ULPCNN 图像自动分割新方法

刘 勍*

为了自动地对图像进行二值分割,提出了一种最大超模糊熵单位链接脉冲耦合神经网络(Unit-Linking PCNN)的图像分割算法。该方法首先对二维超模糊集隶属函数进行了自适应修正,并将其引入到图像超模糊熵概念中;然后从适应图像分割角度考虑,将传统 PCNN 模型改进为具有单调指数上升阈值函数的 ULPCNN 抑制捕获模型;最后把 ULPCNN 与最大超模糊熵判据相结合对图像进行自动分割,并与基于最大香农熵、最小交叉熵及最小模糊熵准则的 ULPCNN 分割方法作了比较。理论分析和实验结果表明,该方法能自动确定迭代次数和选取最佳阈值,对图像目标划分清晰、细节保持较好,改善了图像的分割性能。

0 引言

图像分割是图像分析和处理的重要步骤和技术,其目的是将图像分成具有某种特征差异的不同区域。然而实际图像中往往由于传感器分辨率的限制或者观测目标与周围环境的交叠而使得图像中不同属性的区域混合在一起,很难清晰地将其分开。目前针对图像分割主要有基于边缘的分割[1]、基于阈值的分割[1]、基于聚类的分割[2]及结合特定理论工具的分割等不同方法[3]。而最为简单有效的是阈值分割方法,但如何选取合适阈值,才能进行有效的分割,即分割后二值图像对原图像来说既不产生欠分割,也不引起过分割,这是阈值分割技术的关键。近年来,为控制分割图像中造成的信息损失,在图像分割领域中引入了香农熵(SE)[4]、交叉熵(CE)[5]和模糊熵(FE)[6]等一些客观评判依据,这些不同的分割判据及其与数学形态学[7]、遗传算法[8]、蚁群算法[9]等的结合在图像分割实践中

---

* 作者简介:刘勍(1970—),男,甘肃天水人,天水师范学院教授、博士,主要从事图像信号处理及人工神经网络研究。

虽然取得了一定的效果,但大部分是针对专门图像进行处理的,并且没有充分考虑图像的不精确性和模糊性等固有特性,同时没有完全解决外界干扰因素所引起的图像模糊问题,另外,上述不同的分割方法中为达到最佳分割效果,须有人工干预来确定最佳分割阈值,降低了对图像分割的自动性,大大影响了后续图像分析与理解等处理环节。

脉冲耦合神经网络(pulse coupled neural networks,PCNN)是一种不同于传统人工神经网络的新型神经网络,是依据猫、猴等动物大脑视觉皮层上的同步脉冲发放现象提出的[10],已被广泛地应用于图像平滑、分割等不同的图像处理领域[11~13],并显示了其优越性。然而传统 PCNN 神经元点火状况的复杂性,使得整个网络中脉冲传播行为不便分析与利用,同时该模型在运行过程中不能清晰地反映全局图像阈值,神经网络在图像处理中也只是利用了神经元的点火特性,并且文献[12~13]在 PCNN 迭代过程中只是简单采用香农熵和交叉熵作为图像分割评判准则,没有全面考虑分割图像像素归类的隶属关系,引起对图像目标较大的错误分割,给后续处理带来困难。为此,本文提出了单调指数上升阈值函数的 U-nit – Linking PCNN(ULPCNN)抑制捕获模型;通过引入去除数据本身模糊性和不精确性的超模糊集概念,得到了基于最大超模糊熵准则的 ULPCNN 图像自动分割方法,并与文献[12]提出的基于最大香农熵和文献[13]提出的基于最小交叉熵 PCNN 等其他分割方法做了比较,最后通过理论分析和实验仿真验证了所提方法的可靠性和有效性。

1 超模糊理论

1.1 超模糊集及隶属函数

超模糊集也叫 II 型模糊集,是近些年在经典模糊集合的基础上发展和完善的一种新的模糊集合概念,文献[14~15]对超模糊集合的概念做出了系统的阐述。超模糊集不仅能够去除数据本身的模糊性和不精确性,而且还能够消除经典模糊集隶属函数的不确定性。设 X 为大小为 $M \times N$ 具有 L 灰度级的二维图像数据域,即 $X = \{x_{11}, x_{12}, \cdots, x_{MN}\}$,则 X 上的任一超模糊集 A 的隶属函数 $\mu_A(x_{ij}, u)$,反映了集合中元素 x_{ij} 与这一模糊集 A 的相关程度,其值域为 $0 \leq \mu_A(x_{ij}, u) \leq 1$,此时模糊集 A 可表示为[16]:

$$A = \{((x_{ij}, u), \mu_A(x_{ij}, u)) | \forall x_{ij} \in X, \forall u \in J_{x_{ij}} \subseteq [0,1]\} \qquad (1)$$

其中 $x_{ij} \in X, u \in J_{x_{ij}} \subseteq [0,1]$,$J_{x_{ij}}$ 是[0,1]区间上的一个子区间,A 的离散形式如下:

$$A = \mu_A(x_{11},u)/(x_{11},u) + \mu_A(x_{12},u)/(x_{12},u) + \cdots + \mu_A(x_{MN},u)/(x_{MN},u)$$
$$= \sum_{i=1}^{M}\sum_{j=1}^{N}\mu_A(x_{ij},u)/(x_{ij},u) \tag{2}$$

一般常用不确定性覆盖区域(FOU)来表示超模糊集隶属函数[17],如图1(b)中阴影区域所示。FOU 表示每个隶属函数在阴影区内的值都对应着一个分布函数。即在超模糊集上,某一元素的隶属函数值并不是唯一的,而是一个区间上分布的多个值,这与经典模糊集隶属函数具有单值性是不同的(如图1(a)所示)。为此引入上下限隶属函数来构建每一个元素的FOU(图1(b)所示),便得到一个更具有实际意义超模糊集的表示形式[18]:

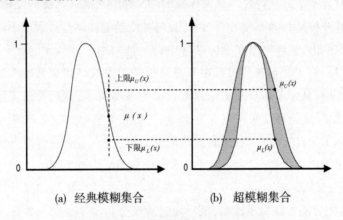

(a) 经典模糊集合　　　(b) 超模糊集合

图1　两种不同模糊集合隶属函数的比较

$$A = \begin{cases} (x_{ij},\mu_U(x_{ij}),\mu_L(x_{ij}))| \ \forall x_{ij} \in X, \\ \mu_L(x_{ij}) \leq \mu(x_{ij}) \leq \mu_U(x_{ij}), \mu \in [0,1] \end{cases} \tag{3}$$

为更好的刻画上下限隶属函数 μ_U 和 μ_L,一般在实际应用中,通常取一对互为倒数的参数来描述它们,其中,隶属函数 μ 的上限 μ_U 和下限 μ_L 分别可表示为:

$$\mu_U(x_{ij}) = [\mu(x_{ij})]^{\varphi_{ij}}, \quad \mu_L(x_{ij}) = [\mu(x_{ij})]^{\frac{1}{\varphi_{ij}}} \tag{4}$$

理论上指数 $\varphi_{ij} \in (0,\infty)$,但是参数的确定会带来一定的困难,文献[19]中选定固定的指数值,并令 $\varphi_{ij} = 1.25$,这不利于不同复杂内容图像分割,对此本文根据图像局部信息,引入式(5)自适应选取像素点 (i,j) 的 φ_{ij} 指数函数值

$$\varphi_{ij} = \frac{\max\limits_{\theta \in \Omega}(x_{(i+\theta)(j+\theta)}) - \min\limits_{\theta \in \Omega}(x_{(i+\theta)(j+\theta)})}{L-1} \tag{5}$$

其中 Ω 为一定宽度的局部窗口。由式(5)可见 φ_{ij} 的取值范围在 $(0,1]$,如果当待处理图像受到不同噪声影响时,局部窗口中最大值和最小值都有可能发生变化,会导致 $\varphi_{ij} > 1$ 的情况,此时可取 $\varphi_{ij} = 1$。由式(4)、(5)的转换关系可知,对超模

糊集的求解和表示可以转换到经典模糊集的情况下来处理,但关键是如何选取隶属函数,图像分割处理中,当采用不同转换隶属函数将会得到不同的超模糊矩阵,文献[15]列出了标准 S 函数、L-R 函数等几种不同的隶属函数,本文从人类视觉对图像灰度响应的非线性关系出发采用 Γ 隶属函数[20],并对其作了简化。Γ 分布概率密度函数的定义表达式如下:

$$f(x) = \frac{\left(\frac{x-\mu'}{\beta}\right)^{\gamma-1} \exp\left(-\frac{x-\mu'}{\beta}\right)}{\Gamma(\gamma)}, \quad x \geq \mu', \gamma, \beta > 0 \qquad (6)$$

其中 γ 是形状参数,μ' 是位置参数,β 是尺度参数及 Γ 是伽马函数。当 $\mu' \neq 0, \beta = 1, \gamma = 1$,式(7)的 Γ 分布退化为式(7)

$$f(x) = \exp[-(x-\mu')], \Gamma(1) = 1 \qquad (7)$$

这样在图像分割中,图像目标区域和背景区域的隶属函数可用式(7)分别去计算,

$$\mu(x_{ij}) = \exp(-\omega|x_{ij}-\mu'_o|), x_{ij} > t, x_{ij} \in X_{object} \qquad (8)$$

$$\mu(x_{ij}) = \exp(-\omega|x_{ij}-\mu'_b|), x_{ij} \leq t, x_{ij} \in X_{background} \qquad (9)$$

其中 ω 为转换系数,它与不同处理图像的像素有关,为了使其具有适应性和鲁棒性,在此选取 $\omega = 1/(x_{max} - x_{min})$,x_{max}, x_{min} 为图像最大灰度和最小灰度值;μ'_o 和 μ'_b 分别是图像目标区域和背景区域的均值,其计算如下

$$\mu'_b = \frac{1}{\sum_{g=0}^{t} h(g)} \sum_{g=0}^{t} g \cdot h(g), \quad \mu'_o = \frac{1}{\sum_{g=t+1}^{L-1} h(g)} \sum_{g=t+1}^{L-1} g \cdot h(g) \qquad (10)$$

其中 g 为像素灰度,$h(g)$ 为图像的灰度统计直方图,t 是假定的分割阈值。

1.2 超模糊熵

熵是信息论中一个最基本并得到广泛应用的概念,它描述了一个概率分布的不确定程度。将熵概念移植到模糊集理论,就得到模糊熵概念,模糊熵描述了一个模糊集的不确定性程度,它是一个下凸函数,模糊熵越大,表示该模糊集越模糊,反之,表示该模糊集越清晰。图像 X 的模糊熵为[18]:

$$FE(\tilde{A}) = \frac{1}{M \cdot N \cdot \ln 2} \sum_{i=1}^{M} \sum_{j=1}^{N} S(\mu_{\tilde{A}}(x_{ij})) \qquad (11)$$

其中 $S(\mu_{\tilde{A}}(x_{ij})) = -\mu_{\tilde{A}}(x_{ij}) \cdot \ln(\mu_{\tilde{A}}(x_{ij})) - (1-\mu_{\tilde{A}}(x_{ij})) \cdot \ln(1-\mu_{\tilde{A}}(x_{ij}))$ (12)

由于模糊熵利用了隶属函数及其补值来反映图像的模糊程度,而模糊集中隶属函数是一个确定的值,在图像分割实践中不可能把某个像素点精确地划分到背景或目标区域,这只能从理论的角度,估计该像素点隶属于背景或目标区域的可能范围。而超模糊集隶属函数具有多值性,体现一个数据属于某个范围的程度,据此我们把模糊熵进行推广,引入超模糊熵(UFE)概念,使其能更好的满足图像

分割的要求,并具有更广泛的实际意义,其数学表示为

$$UFE(A) = \frac{\sum_{i=1}^{M}\sum_{j=1}^{N}(-\mu_U(x_{ij}) \cdot \ln\mu_U(x_{ij}) - \mu_L(x_{ij}) \cdot \ln\mu_L(x_{ij}))}{M \cdot N \cdot \ln 2} \tag{13}$$

其中 $\mu_U(x_{ij})$ 和 $\mu_L(x_{ij})$ 分别是隶属函数 $\mu(x_{ij})$ 的上限和下限,它们的取值与图像分割阈值有关,当选定不同的分割阈值时,会得到不同的超模糊熵,即超模糊熵 $UFE(A)$ 是分割阈值 t 的函数,其大小又将直接影响分割阈值的选取,而在式(13)中 $\mu_U(x_{ij})$ 和 $\mu_L(x_{ij})$ 采用了互为倒数的指数 φ_t 调节,这样使形成的 $UFE(A)$ 函数具有上凸性,只有当超模糊熵 $UFE(A)$ 达到最大时,确定的阈值为最佳分割阈值,此时可得到最佳二值分割图像。

2 改进 ULPCNN 模型及图像分割

PCNN 也称为第三代人工神经网络,构成 PCNN 的每一神经元由接收、调制和脉冲产生 3 部分组成,传统 PCNN 模型人工设置参数多、适应性能差且只利用了生物神经元的激活特性及其阈值指数衰减特性,虽然按指数衰减的阈值变化规律符合人眼对灰度响应的非线性要求,但是图像分割的目的是为了区分目标和背景,或不同的目标区,在目标、背景或不同目标间,由于像素灰度的相似性较差,因而给分割带来一定的困难。为了更好地适应图像处理应用的要求,从图像分割的目标及精度出发,在不严格符合真实生物神经元性质前提下,降低神经元链接通道信号的复杂性,利用神经元抑制捕获特性,并采用传统阈值分割搜索策略(即采用由小到大单调增长的阈值函数),对传统 PCNN 改进简化为 ULPCNN 模型,其离散数学方程描述如下:

$$F_{ij}[n] = x_{ij} \tag{14}$$

$$L_{ij}[n] = \begin{cases} 1, & \sum_{k,l \in W} Y_{kl}[n] > 0 \\ 0, & otherwise \end{cases} \tag{15}$$

$$U_{ij}[n] = F_{ij}[n](1 + \lambda L_{ij}[n]) \tag{16}$$

$$Y_{ij}[n] = \begin{cases} 1, & (U_{ij}[n] \geq T_{ij}[n]) \\ 0, & (U_{ij}[n] < T_{ij}[n]) \end{cases} \tag{17}$$

$$t_{ij}[n] = \begin{cases} t_0 \cdot \exp(-\alpha_t/n), & (Y_{ij}[n-1] = 1) \\ T_0, & (Y_{ij}[n-1] = 0) \end{cases} \tag{18}$$

其中 ij 下标为神经元的标号,n 为迭代次数,x_{ij}、$F_{ij}[n]$、$L_{ij}[n]$、$U_{ij}[n]$、$t_{ij}[n]$

分别为神经元的外部刺激、第 ij 个神经元第 n 次反馈输入、连接输入、内部活动项和动态阈值,W 为链接权矩阵,λ 为链接强度,t_0 为阈值幅度常数,自适应选取为图像像素灰度最大值 $x_{\max}$,α_t,T_0 为待选常数,$Y_{ij}[n]$ 是 PCNN 的二值输出。

PCNN 进行图像处理时,首先将一个二维 PCNN 网络的 $M\times N$ 个神经元分别与二维输入图像的 $M\times N$ 个像素相对应,将每个像素点的灰度值输入到对应神经元的连接输入,同时每个神经元的输出与其邻域中其他神经元输入相连。在本文的分割过程中,每个神经元只能激活一次,即在第 1 次迭代时,首先给出一全局零阈值,让所有像素都激活,并按照式(18)产生下一次迭代的阈值,然后依照以上关系进行迭代,当内部连接矩阵 W 所在的邻域有灰度值相近的像素存在时,且其中某一像素灰度小于输入阈值时,由其抑制产生的脉动输出依次传递将会引起附近其他类似灰度像素对应神经元的抑制,而产生脉动输出序列 $Y[n]$。由离散时间第 n 次的输出序列 $Y[n]$ 构成的二值图像就是 PCNN 输出分割图像。

运用 ULPCNN 可以分割出不同迭代次数下不同层次的二值图像,但如何确定最优分割结果和选择最佳分割阈值,以达到图像自动分割的目的,本文引入最大超模糊熵准则作为判优凭据,即在给定一迭代次数 n_0 后,对输入图像进行 ULPCNN 处理,并计算 $t[n]$ 及 $UFE(A)$ 值,最后求取使 $UFE(A)$ 最大时 $UFE\max$ 的迭代次数 $n_{\max}^{UFE}$ 及最佳阈值 t_{opt}^{UFE},由其对应的输出 $Y[n_{\max}^{UFE}]$ 构成的二值图像,即为最佳分割输出图像。

3 实验结果与分析

基于 MATLAB 仿真环境,对大量不同灰度图像进行基于 UFE 最大准则改进 ULPCNN 分割算法实验研究,为验证所提算法的可适用性,特选取 256×256 灰度级为 256 目标复杂、细节丰富的 Lena 图像、目标和背景对比变化较小医学细胞图像和 Rice 图像的实验结果,并与基于 CE 最小准则[12]、SE 最大准则[13]及 FE 最小准则的 ULPCNN 分割方法进行比较,实验中 Ω 选为 3×3 的局部窗口,$\lambda=0.1$,$\alpha_t=13$,$t_0=255$,$T_0=1000$,图像分割结果如图 2~4 所示。

为了客观评价图像分割效果,本文采用区域内部均匀性 UM 和区域对比度 CM[21] 图像分割评价标准分别对四种不同准则的 ULPCNN 分割结果进行评价,并做出对比,结果如表 1 所示,其中

$$UM = 1 - \frac{1}{C}\sum_s\left\{\sum_{ij\in R_s}\left[x_{ij} - \frac{1}{C_s}\sum_{ij\in R_s}x_{ij}\right]^2\right\} \qquad (19)$$

$$CM = \frac{|\mu'_0 - \mu'_b|}{\mu'_0 + \mu'_b} \qquad (20)$$

C、C_s 为归一化系数,μ'_0 和 μ'_b 分别为目标和背景的灰度均值,UM 和 CM 值越大,分割效果越好,算法性能也越好。

从实验图 2～图 4 可以看出在 ULPCNN 迭代分割过程中,运用四种准则均能自动获取各自分割的最佳二值图像。但主观视觉效果明显看出 UFE 最大分割准则要优于其他三种分割准则。具体体现在:Lena 图像中目标划分清晰(如面部嘴、鼻子、眼睛、帽子的横纹以及梳妆镜的隔框等小目标部分)、细节保持较好(如帽子上的饰物和肩部头发),而基于 CE 和 FE 最小准则分割在这些区域则产生了欠分割,没有分割出较小的目标区域,基于 SE 最大分割准则形成了过分割,许多小区域被背景掩盖而没有划分出来,甚至还产生了误分割(如肩部及梳妆镜右上角等区域);医学细胞图像中基于 UFE 最大分割准则较好的把目标提取与分离出来,而其它三种准则的分割方法对图像右下角的细胞分离不彻底或完全没有从背景中分割出来,同时误把部分背景分成目标或产生噪点,出现了过分割现象,且细胞内部高亮度小区域划分不清晰;Rice 图像中 CE 和 SE 准则产生了过分割,而 FE 准则形成了欠分割,有部分目标没有分割出来。

(a) 原始图像　　(b) CE最小准则　　(c) SE最大准则　　(d) FE最小准则　　(d) UFE最大准则

图 2　对 Lena 图像基于四种不同熵准则 ULPCNN 分割结果

 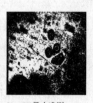 

(a) 原始图像　　(b) CE最小准则　　(c) SE最大准则　　(d) FE最小准则　　(e) UFE最大准则

图 3　对医学细胞图像基于四种不同熵准则的 ULPCNN 分割结果

在几种分割判优准则中,香农熵强调系统内部的均匀性,其信息量的大小,仅仅依赖于结果中 0 和 1 所占的比例,因此,如果分割结果中 0 和 1 的概率相等,则会产生香农熵的最大值,但图像的复杂性并不能保证目标像素和背景像素的个数大致相等;而交叉熵反映两个系统间信息量的差异,即分割前后两幅图像信息量的差,也就是分割后二值图像取代原图像时,其信息量变化的期望值,但当图像中

(a) 原始图像　　(b) CE最小准则　　(c) SE最大准则　　(d) FE最小准则　　(d) UFE最大准则

图 4　对 Rice 图像基于四种不同熵准则的 ULPCNN 分割结果

目标灰度变化较小或产生一定失真畸变时,其交叉熵几乎不变,不利于分割阈值的衡量与测定;模糊熵考察的是图像系统中某像素精确隶属于目标或背景的可能程度,其值越大说明图像目标分离越不清晰,但图像小目标区域或目标与背景交界区域像素划分具有一定的区间模糊性,不是精确归类的,所以模糊熵准则的衡量就显得有些苍白无力;而超模糊熵中隶属函数的多值性正好弥补了模糊熵中隶属函数的不足,可以比较准确的确定某个像素在多大程度隶属于背景或目标区域的范围,能有效地去除数据本身的模糊性和不精确性,具有广泛的适用性,因而可以更准确地评判分割的优劣。这可以从表 1 中不同图像 ULPCNN 分割的区域均匀性和区域对比度客观评价指标数据反映出其优劣来。

表 1　基于四种不同熵准则的 ULPCNN 分割实验数据

	准则	熵值	迭代次数	分割阈值	区域均匀性	区域对比度	执行时间(s)
Lena 图像	CE 最小	0.2732	11	78	0.9854	0.4650	18.1533
	SE 最大	0.9993	14	100	0.9858	0.4312	20.2622
	FE 最小	0.4240	13	94	0.9862	0.4464	22.4549
	UFE 最大	0.3164	12	86	0.9870	0.4723	24.9756
Medicinal cell 图像	CE 最小	0.3583	33	157	0.9918	0.1351	54.7173
	SE 最大	0.9992	30	150	0.9929	0.1434	43.6530
	FE 最小	0.2852	27	141	0.9938	0.1637	46.8477
	UFE 最大	0.3930	15	110	0.9941	0.2956	27.0865
Rice 图像	CE 最小	0.4098	14	100	0.9877	0.2944	24.4160
	SE 最大	0.9892	16	113	0.9914	0.3022	24.6561
	FE 最小	0.3824	29	163	0.9904	0.3157	52.8090
	UFE 最大	0.3834	19	128	0.9928	0.3180	36.0751

传统 PCNN 模型的阈值是反复变化的,长时间(或短时间)的衰降之后必然会有由于神经元激活而造成的突然上升,之后又是衰降,然后又是突然上升……

这种机理使得 PCNN 处理后的大量信息蕴含在神经元激活周期(频率)或者激活相位中,而直接的二值输出图像却并不包含全部的信息。而本文选取严格单调增加指数阈值函数,是随时间收敛于 t_0 的,这个特点使得灰度相近的神经元一定能够被分成一类,从实验结果图 2~4 可以看到,利用改进 ULPCNN 模型的抑制捕获特性,可以执行有效的图像分割,同时由表 2 可以得到改进 ULPCNN 模型比传统 PCNN 在使用较少迭代时间内产生最佳分割结果,且其分割的区域均匀性和区域对比度均好于传统 PCNN,充分说明了抑制捕获指数上升阈值 ULPCNN 模型是对传统 PCNN 模型在图像处理领域应用中有益补充和拓展。

表 2 基于 UFE 准则采用两种 PCNN 模型对医学细胞图像的分割实验数据

	最佳迭代次数	最大熵值	区域均匀性	区域对比度
传统 PCNN	25	0.3875	0.9929	0.1683
改进 ULPCNN	15	0.3930	0.9941	0.2956

图 5 $\varphi_{ij}=1.25$ 时三幅图像 UFE 最大准则 ULPCNN 分割结果

表 3 $\varphi_{ij}=1.25$ 时 ULPCNN 分割数据

图像	最佳迭代次数	UFE 最大值
Lena	6	0.2766
cell	6	0.2267
Rice	7	0.3685

4 结论

本文提出了一种基于超模糊熵的 ULPCNN 图像自动分割新方法,该方法在分析和探讨超模糊集理论的基础上,对隶属函数进行了自适应修正,并将其引入到二维图像超模糊熵概念中;然后从适合图像分割的角度出发,将传统 PCNN 模型改进简化为指数上升动态阈值函数的单位链接 PCNN 抑制捕获模型;最后将其与最大超模糊熵判据相结合对图像进行最佳自动分割,并与基于最大香农熵、最小交叉熵及最小模糊熵准则的 ULPCNN 分割方法作了比较。理论分析和实验结果表明,本文算法对图像分割精度高、目标划分清晰、细节保持较好、分割质量接近最佳。

参考文献

[1]黄长专,王彪,杨忠.图像分割方法研究[J].计算机技术与发展,2009,19(6):76-79.

HUANG Chang-zhuan,WANG Biao,YANG Zhong. A Study on Image Segmentation Technique[J]. Computer Technology and Development,2009,19(6):76-79.

[2]刘德连,王博,张建奇.期望最大化聚类的高光谱亚像素目标检测[J].西安电子科技大学学报,2009,36(3):512-516.

LIU De-lian,WANG Bo,ZHANG Jian-qi. Hyperspectral subpixel target detection approach based on expectation-maximization cluster[J]. Journal of Xidian University,2009,36(3):512-516.

[3]孙强,焦李成,侯彪.基于子波域自适应融合 HMTseg 算法的遥感图像分割[J].西安电子科技大学学报,2007,34(6):853-858.

SUN Qiang,JIAO Li-cheng,HOUBiao. Remotely sensed image segmentation based on the wavelet-domain
HMTseg algorithm with adaptive fusion mechanism[J]. 西安电子科技大学学报,2007,34(6):853-858.

[4]Pal N R,Pal S K. Entropic thresholding [J]. Signal Processing,1989,16(2):97~108.

[5]AL-Osaimi,Ghada. EL-Zaart,Ali. Minimum cross entropy thresholding for SAR images [A]. 2008 3rd International Conference on Information and Communication Technologis:From Theory to Applications[C]. Damascus,Syrian Arab Republic,2008:4530-4535.

[6]Li Linyi,li Deren. Fuzzy entropy image segmentation based on particle swarm optimization [J]. Progress in Natural Science,2008,18(9):1167-1172.

[7]Yong Yang,Shuying Huang. Retinal Image Mosaic Base on Genetic Algorithm and Automated Blood Vessel Extracting Approach [A]. Proceedings of the 7th World Congress on Intelligent Control and Automation[C]. Chongqing,China,2008,7751-7756.

[8]Mo Xiao-qi,Wang Yao-nan. Entropic Methods for Segmentation of Human Chromosome Images based on Genetic Algorithm[J]. Journal of System Simulation. 2006,18(7):1921-1925.

[9]Xiaohong Shen,Yulin Zhang,Fangzhen Li. An Improved Two-Dimensional Entropic Thresholding Method Based on Ant Colony Genetic Algorithm[A]. 2009 Global Congress on Intelligent Systems[C]. Xiamen,China. 2009,163-167.

[10] R Eckhorn, H J Reitboeck, M Arndtetal. Feature linking via synchronization among distributed assemblies: simulation of results from cat cortex [J]. Neural Computation, 1990, 2(3): 293-307.

[11] 刘勃,马义德. 一种基于PCNN赋时矩阵的图像去噪新算法[J]. 电子与信息学报, 2008, 30(8): 1869-1873.
Liu Qing, Ma Yi-de. A New algorithm for Noise Reducing of Image Based on PCNN Time Matrix[J]. Journal of Electronics & Information Technology, 2008, 30(8): 1869-1873.

[12] 马义德,戴若兰,李廉. 一种基于脉冲耦合神经网络和图像熵的自动图像分割方法[J]. 通信学报, 2002, 23(1): 46~51
Ma Yide, Dai Ruolan, Li Lian. Automated Image Segmentation Using Pulse Coupled Neural Networks and Image Entropy [J] Journal of China Institute of Communications, 2002, 23(1): 46~51.

[13] Ma Yi-de, Liu Qing, Qian Zhi-Bai. Automated image segmentation using improved PCNN model based on cross-entropy [A]. Proceedings of 2004 International Symposium on Intelligent Multimedia, Video and Speech Processing. [C]. Hong Kong, China, 2004: 743-746.

[14] J M Mendel, R I Bob John. Type-2 fuzzy sets made simple[J]. IEEE Trans. Fuzzy Systems(S1063-6706), 2002, 10(2): 117-127.

[15] Nilesh N-Karnik, Jerry M-Mendel. Operations on type-2 fuzzy sets[J]. Fuzzy Sets and Systems(S0165-0114), 2001, 122(2): 327-348.

[16] Hamid R. Tizhoosh. Type II Fuzzy Image Segmentation[J]. Fuzzy Sets and Their Extensions: Representation, Aggregation and Models. Springer. 2008, 607-619.

[17] J. M. Mendel and R. I. Bob John, Type-2 Fuzzy Sets Made Simple[J]. IEEE Transactions on Fuzzy Systems, 2002, 10(2): 117-127.

[18] H. R. Tizhoosh, Image Thresholding Using Type II Fuzzy Sets[J]. Pattern Recognition, 2005, 38: 2363-2372.

[19] 片兆宇,高立群,吴建华,等. 基于超模糊集的多属性图像阈值分割算法[J]. 系统仿真学报, 2007, 19(19): 4434-4439.
PIAN Zhao-yu, GAO Li-qun, WU Jian-hua, et al. Image Thresholding and Simulation Using Multi-properties Based on Ultra-fuzzy Sets[J]. Journal of System Simulation, 2007, 19(19): 4434-4439.

[20] Chaira T, Ray, A. K. Segmentation using fuzzy divergence[J]. Pattern Recognition Lett, 2003, 12(24): 1837-1844.

[21] Cardoso J S, Luls C R. Toward a generic evaluation of image segmentation. IEEE Transactions on Image Processing, 2005, 14(11): 1773-1782.

注:本文曾发表在2010年10月《西安电子科技大学学报》第5期上。

一种频率域提高 Radon 变换分辨率的方法

刘保童*

Radon 变换是压制相干噪音，波场分离的重要方法之一。算子假频和端点效应是该方法在数值计算中需要始终关注的两个重要问题，不断地改进和完善变换的具体算法，抑制和最大化地减少算子假频和端点效应，才能不断地提高变换的分辨率和质量，促进和发展 Radon 变换的有效应用。针对在数值计算中应该关注解决的问题，在用反演理论对变换的非唯一性进行分析的基础上，本文对常规的最小平方 Radon 变换方法做了改进，给出一种频率域 Radon 变换方法，可有效地压制端点效应，提高了变换域的分辨率。数值计算试验表明了该方法的有效性。

0 引言

Radon 变换最早是由奥地利数学家 J. Radon 于 1917 年提出的[1]，为物理学、医学、天文学、分子生物学、材料科学、光学、核磁共振、无损检测技术和地球物理学中的一大类图像重建（层析）问题提供了统一的数学基础。在 20 世纪 70 年代中期，由美国斯坦福大学以 Claerbout 为首的地球物理小组将 $\tau-p$ 变换引入到地震勘探中，提高了人们对 $\tau-p$ 变换的研究兴趣[2,3]。Thorson 和 Claerbout(1985) 首先给出了 Radon 变换的最小平方公式和随机反演公式[4]，Beylkin(1987) 提出了离散 Radon 变换(DRT)[5]。Beylkin(1987)、Kostov(1990) 分别研究了计算 Radon 变换的最小平方方法，Foster 和 Mosher(1992) 将这种技术用于压制多次波[6]，Zhou 和 Greenhalgh(1994) 分析讨论了最小平方线性 $\tau-p$ 变换与抛物型 $\tau-p$ 变换的分辨率问题[1]，Sacchi 和 Ulrych(1995) 提出了高分辨率抛物型 Radon 变换[7]。吴律对 Radon 变换的一般原理和应用做了比较系统的论述[2]，牛滨华等人深入研究了 Radon 变换并提出多项式 Radon 变换($\tau-pq$ 变换)[3]，还有沈操、黄新武、魏

---

* 作者简介：刘保童(1965—   )，男，甘肃天水人，天水师范学院教授、博士，主要从事数字信息处理与物理学研究。

修成、朱生旺、李远钦等许多作者也都对 Radon 变换做了很好的研究[2,8-10]。针对地面地震勘探中普遍存在的多次波压制、噪声衰减、波场分离等问题,近年来国内外一直在对提高 Radon 变换的分辨率进行研究,成为一个热点,一般都是采用迭代稀疏反演方法,Daniel Trad et al. (2003) 对稀疏 Radon 变换做了总结分析[11]。在地面叠前、叠后线性相干噪声压制以及 VSP 波场分离中,研究如何提高线性 Radon 变换的分辨率显得非常重要,而改善抛物线 Radon 变换的分辨率可以更为有效地压制多次波[12]。本文在用反演理论对 Radon 变换的非唯一性进行分析的基础上,给出一种非迭代方法,能有效提高变换分辨率。通过数值计算试验显示了该方法的有效性。

1 Radon 变换的原理及数值计算时存在的问题

二维 Radon 变换的一般形式为

$$m(\tau,p) = \iint d(x,t)\delta(t - \theta(\tau,p,x))dtdx \tag{1}$$

对倾斜叠加,时差曲线 $\theta(\tau,p,x) = \tau + px$,(1)式简化为

$$m(\tau,p) = \int_{-\infty}^{+\infty} d(x,\tau + px)dx \tag{2}$$

对应的反变换(反倾斜叠加)为

$$d(x,t) = \int_{-\infty}^{+\infty} m(t - px,p)dp \tag{3}$$

取(3)式的有限离散形式

$$d(x_k,t_l) = \sum_{j=1}^{J} m(t_l - p_j x_k, p_j) \tag{4}$$

对(4)式两边做傅里叶变换到频率域,得到线性方程组

$$D = GM \tag{5}$$

由(2)式类似可得

$$M = G^H D \tag{6}$$

其中 $G = (\exp(-i\omega p_j x_k))_{K \times J}$;$D(x_k,\omega)$ 为输入记录的傅氏变换;$M(\omega,p_j)$ 为 τ-p 谱的傅氏变换;$l = 1,2,\cdots,L$;$j = 1,2,\cdots,J$;$k = 1,2,\cdots,K$;H 表示共轭转置。

为提高变换分辨率,先定义反变换(5)式,然后导出正变换[1,3,6,8],而由(5)式求 M 是一个线性反演问题,存在解的非唯一性,即对数据空间中已知的观测结果 D,在模型空间中存在多个可能的 M,它们都能表示 D。非唯一性从何而来?被誉为反演理论之父的 Backus 和 Gilbert 认为:"来源于观测数据的数目并非无限,以及观测数据具有误差。仅此而已!"

由于 G 是一个 $K \times J$ 的矩阵,而方程(5)一般是欠定方程组,G 的逆应该用 Moore – Penrose 广义逆来求解。(5)式的最小范数解[4]

$$M = G^H (GG^H + \lambda^2 I)^{-1} D \tag{7}$$

式中 λ 为阻尼因子,一般取 0.1~1 即可。(7)式就是反问题(5)的阻尼最小平方解,$(GG^H + \lambda^2 I)^{-1}$ 等价于一个反褶积算子,它提高了变换域的分辨率[1,3,6,7,13]。但(7)式给出的这个反演解还不够理想,分辨率仍不能满足信噪分离的要求,因此,寻求反问题(5)式的具有更高分辨率的解显得非常重要。

Radon 变换法去噪或波场分离寄希望于在变换域中信号与噪声或不同类型的波彼此能够分离。(2)式表明,线性 Radon 变换把时 – 空域($t - x$ 域)中的直线映射为截距时间 – 慢度域($\tau - p$ 域)中的点,然而,在数值计算实现中,在变换域往往会出现能量扩散,而使去噪或波场分离效果不好。影响变换质量的原因有两个:算子假频和端点效应,它们有时被称为离散 Radon 变换的缺陷[13]。这些人工干扰不仅会在 Radon 变换的时不变形式(线性,抛物线)所得到的图像中出现,也在广义时空变 Radon 变换(如 Kirchhoff 偏移)的结果中出现。有限的空间采样会引起假频干扰,而有限的空间排列孔径(观测窗)产生端点效应,限制了 Radon 变换的分辨率。算子假频干扰的消除是一个相对来说直截了当和好理解的问题,考虑输入信号(被积函数)的带限性质,按照采样定理选择合适的扫描范围 [p_{min}, p_{max}] 和 Δp 即可避免[2,3];但端点效应干扰提出了一个更加困难的问题,它降低了变换域的分辨率。实现 Radon 变换以得到不受这些限制的分辨率是一个不适定的反问题,在形式上可与稀疏脉冲反演(例如 Oldenburg et al.,1983)相比,因而只能通过加先验约束来实现,且一般都要迭代[7,11,14]。本文下面给出一种非迭代的反演解,可有效地提高变换分辨率。

2 一种提高 Radon 变换分辨率的方法

由前面的分析可知,反问题(5)是一个约束的欠定最小平方问题,因此,按照广义最小平方理论[15],本文给出(5)的一个解

$$M(\omega_n) = WG^H (GWG^H + \lambda^2 I)^{-1} D(\omega_n) \tag{8}$$

式中 ω_n 为第 n 个角频率成分,$D(\omega_n) = (D(x_1,\omega_n), D(x_2,\omega_n), \cdots, D(x_K,\omega_n))^T$ 为要变换的单色波场,$M(\omega_n) = (M(\omega_n,p_1), M(\omega_n,p_2), \cdots, M(\omega_n,p_J))^T$ 为反演得到的变换结果,$G(\omega_n)$ 为(5)中的变换矩阵,$W(\omega_n)$ 是一个实对角正定约束矩阵,其对角元素为

$$w_{jj}(\omega_n) = \| M(\omega_{n-1}, p_j) \| \quad (j = 1,2,\cdots,J) \tag{9}$$

这种非迭代的直接处理方法在没有同相轴斜率先验信息的情况下,能够将输入数

据的单色波分解聚焦到其最大的谱成分上。如果在(1)式中取时差曲线 $\theta(\tau,q,x) = \tau + qx^2$,则该方法可进行抛物线 Radon 变换($\tau-q$ 变换),这时的 q 为抛物线的曲率。

3 数值计算试验

图 1 是一个水平界面的合成地震记录,共 19 道,双边接收,炮检距范围[-90m,+90m],道距10m,取 $\Delta p = 0.00002$,p 的扫描范围为[-0.004,+0.004],普通 Radon 变换的结果显示在图 2 中,既有算子假频也有断点效应。为避免假频,扫描范围 $[p_{\min}, p_{\max}]$ 和 Δp 的选择要符合采样定理[2]。例如,若选择 p 的扫描范围为[-0.002,+0.002],避免了假频,但端点效应依然未变(图 3)。图 4 是用(7)式变换的结果,由于在 p 方向做了反褶积,分辨率比图 3 有了明显的改进,但并不理想。图 5 是用本文介绍的方法得到的变换结果,分辨率的提高是显著的。图 6 是图 5 的重建结果,完全恢复了原始记录。图 7 是含有倾斜相干噪声的合成记录,最小炮检距 0m,最大炮检距 580m,道间距 20m,图 8 是采用本文的方法得到的 $\tau-p$ 谱,p 的扫描范围[-0.002,+0.004],$\Delta p = 0.00001$,挖去左下方和右上方的两个能量团(对应两个倾斜干扰)后重建的结果显示在图 9 中,由于在变换域中能量聚焦好,倾斜相干噪声消除后,与倾斜干扰相交叉处的有效波没有损失和畸变(图中箭头所指)。图 10 对抛物线 Radon 变换做了数值试验,该图中,(a)是一个经 NMO 校正的合成 CMP 道集,(b)是据普通 $\tau-q$ 变换的谱(d)消除多次波后重构的结果,可见,多次波未去除干净且有效波受到损伤,(c)是据用本文方法得到的 $\tau-q$ 谱(e)消除多次波后重构的结果,克服了(b)中的不足。这里多次波的消除是采用将(d)和(e)中粗短划线右面的部分切除的做法。图 11 是一个实际 VSP 记录,图 12 是用文中介绍的线性 Radon 变换方法分离的上行波。

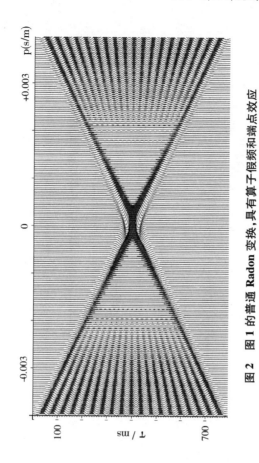

图 2 图 1 的普通 Radon 变换,具有算子假频和端点效应

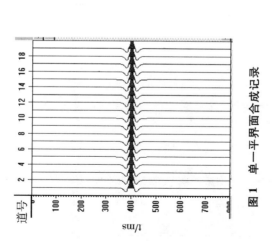

图 1 单一平界面合成记录

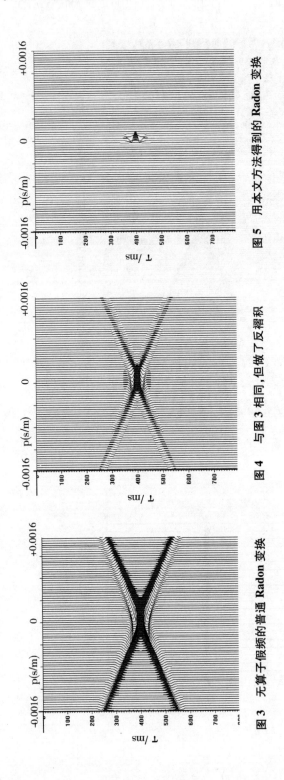

图 3 无算子限频的普通 Radon 变换

图 4 与图 3 相同,但做了反褶积

图 5 用本文方法得到的 Radon 变换

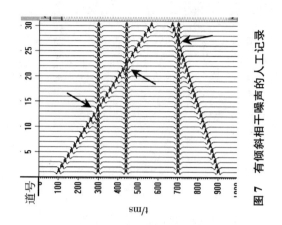

图 7 有倾斜相干噪声的人工记录

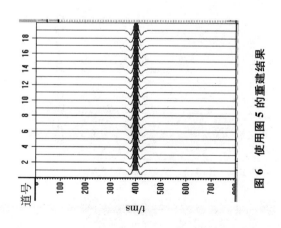

图 6 使用图 5 的重建结果

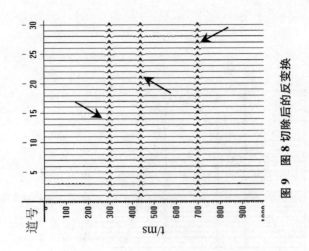

图 9 图 8 切除后的反变换

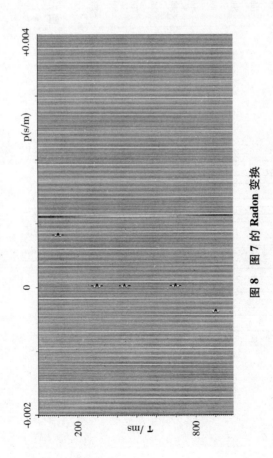

图 8 图 7 的 Radon 变换

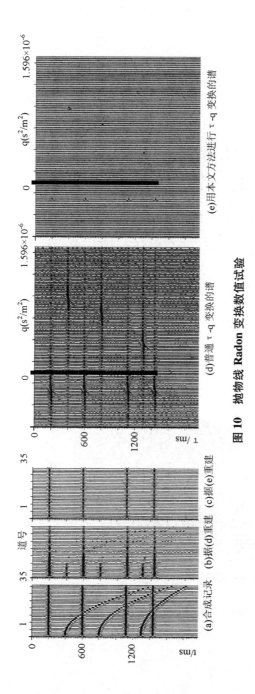

图 10 抛物线 Radon 变换数值试验

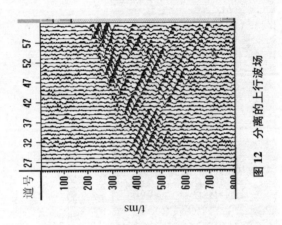

图 12 分离的上行波场

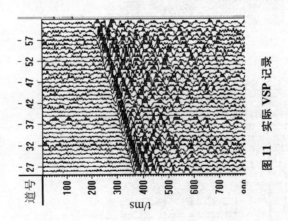

图 11 实际 VSP 记录

4 结束语

在 Radon 变换的数值计算中,应不断地改进和完善变换的具体算法,抑制和最大化地减少算子假频和端点效应,才能不断地提高变换的分辨率和质量,促进和发展 Radon 变换的有效应用。多年来,国内外许多学者都为此做了大量的研究,也取得了许多有效的成果。本文在前人研究工作的基础上,对常规的最小平方 Radon 变换方法做了改进,所给出的一个频率域约束最小平方反演解可压制端点效应,显著提高变换分辨率。

参考文献

[1] BinZhong Zhou and Stewart A Greenhalgh. Linear and parabolic $\tau - p$ transforms revisited [J]. Geophysics,1994,59(7):1133 - 1149.

[2] 吴律. $\tau - p$ 变换及应用[M].北京:石油工业出版社,1993.

[3] 牛滨华,孙春岩,张中杰,等.多项式 Radon 变换.地球物理学报[J],2001,44(2):263 - 271.

[4] Thorson J R,Claerbout J F. Velocity - stack and slant - stack stochastic inversion[J]. Geophysics,1985,50(12):2727 - 2741.

[5] Beylkin G. Discrete Radon transform[J]. IEEE Trans. , Acoust. , speech, and Signal Proc. ,1987,ASSP - 35(2):162 - 172.

[6] Foster D J,Mosher C C. Suppression of multiple reflections using the Radon transform[J]. Geophysics,1992,57(3):386 - 395.

[7] Sacchi M D,Ulrych T J. High resolution velocity gathers and offset - space reconstruction [J]. Geophysics,1995,60(4):1169 - 1177.

[8] 沈操,牛滨华,余钦范. Radon 变换的 MATLAB 实现[J].物探化探计算技术,2000,22(4):346 - 350.

[9] 黄新武,吴律,牛滨华,等.抛物线 Radon 变换中的参数采样与假频[J].石油大学学报(自然科学版),2003,27(2):27 - 31.

[10] 朱生旺,魏修成,李峰,等.用抛物线 Radon 变换稀疏解分离和压制多次波[J].石油地球物理勘探,2002,37(2):110 - 115.

[11] Daniel Trad, Tadeusz Ulrych, and Mauricio Sacchi. Latest views of the sparse Radon transform. Geophysics,2003,68(1):386 ~ 399

[12] Yanghua Wang. Multiple attenuation:coping with the spatial truncation effect in the Radon transform domain[J]. Geophysical Prospecting,2003,51:75 - 87.

[13] Marfurt K J,Schneider R V,and Mueller M C. Pitfalls of using conventional and discrete Radon transforms on poorly sampled data[J]. Geophysics,1996,61(5):1467 - 1482.

[14] Cary P W. The simplest discrete Radon transform[C]. 69th Ann. Internat. Mtg. ,Soc. Expl. Geophys. ,1998.

[15] Tarantola A. Inverse problem theory:Methods for data fitting and model parameter estimation[M]:Elsevier Science Publ. ,1987.

[16] 田玉仙. 傅立叶变换及其在夫朗和费圆孔衍射中的应用[J]. 西安科技大学学报, 2005,25(1):106-108.

注:本文曾发表在2006年1月《西安科技大学》第1期上。

非均匀 DFT 频谱泄漏抑制方法研究

刘保童*

针对不规则采样信号的谱估计问题,提出一种非均匀离散傅里叶变换频谱泄漏抑制方法,通过迭代非线性估计实现非均匀采样信号离散傅里叶变换的计算。实际计算试验结果显示,这种方法能有效地抑制非均匀离散傅里叶变换结果中的频谱泄漏,提高 DFT 频谱的分辨率。

离散傅里叶变换(Discrete Fourier Transform,DFT)在各种数字信号处理的算法中起着核心作用,其应用已遍及各个科学技术领域[1,2]。因为用 DFT 对信号进行频谱分析时,要求信号必须是有限长的离散信号,所以必须对被分析信号进行加窗截断,因此,频谱泄漏是不可避免的。泄漏现象是离散傅立叶变换所固有的,对于一个周期函数,当截取的时间间隔不是周期函数周期数的整数倍时,则会引起时域波形的间断,相应地在频域产生了旁瓣。在离散傅里叶分析中,频谱能量泄漏定义为:由于窗函数引入的谱平滑作用,使得在一个频率处的分量泄漏到相邻的频率分量中去[2]。泄漏使得 DFT 结果仅仅是原信号真实频谱的一个近似,尽管能量泄漏也能起到克服栅栏效应的积极作用,但它的存在有时会掩盖真实谱中的较小的谱峰,这是影响谱估计精度的重要因素,泄漏也会造成混叠,因此,虽然我们没有办法完全消除泄漏,但在实际中应尽可能设法减小 DFT 频谱能量的泄漏。关于均匀采样信号 DFT 频谱泄漏的抑制方法,研究比较成熟,已有多种方法[3-6],其中最常用的是加窗法[7]。非均匀采样信号的 DFT 也存在频谱能量的泄漏,而且往往比较严重,如何抑制非均匀 DFT 频谱能量的泄漏,未见公开发表的文献。本文提出一种抑制非均匀 DFT 频谱泄漏的方法,可有效地减小频谱能量的泄漏,提高 DFT 的分辨率。

---

* 作者简介:刘保童(1965—),男,甘肃天水人,天水师范学院教授、博士,主要从事数字信息处理与物理学研究。

本文行文如下。首先基于线性反演理论,构造目标函数,采用柯西(Cauchy)分布施加的正则化项,使目标函数最小化获得正则化的唯一解,它是计算非均匀离散傅里叶变换(Nonuniform Discrete Fourier Transform, NDFT)频谱的基础。然后,给出了频谱的迭代非线性估计式。进行了数值计算试验,并给出了缺失数据重建的一个应用实例,具体的计算结果表明,这种方法能有效地抑制非均匀离散傅里叶变换结果中的频谱泄漏。

1 基于线性反演理论的非均匀DFT频谱估计

定义连续正向空间傅里叶变换为

$$P(k_x) = \int_{-\infty}^{+\infty} p(x) e^{jk_x x} dx \tag{1}$$

式中 x 是空间变量,k_x 是圆波数。反变换由下式给出

$$p(x) = \frac{1}{2\pi} \int_{-\infty}^{+\infty} P(k_x) e^{-jk_x x} dk_x \tag{2}$$

时间傅里叶变换对用类似的方式定义,只不过指数所用符号的约定不同。对于沿 x 规则采样的带限数据,众所周知,与(1)式有关的离散傅里叶变换(DFT)为

$$P(k_x) = \sum_{n=0}^{N-1} p(n\Delta x) e^{jk_x n\Delta x} \Delta x \tag{3}$$

其中 Δx 的选择应足够小以避免空间傅里叶域假频。在不规则采样的情况下,一种获得傅里叶域数据的直接方法是使用黎曼和(the Riemann sum),(1)式中的积分用一个与实际采样位置 $(x_0, x_1, \cdots, x_{N-1})$ 对应的和来代替:

$$P(k_x) = \sum_{n=0}^{N-1} p(x_n) e^{jk_x x_n} \Delta x_n \tag{4}$$

式中 Δx_n 定义为

$$\Delta x_n = l_{n+1} - l_n = \frac{x_{n+1} - x_{n-1}}{2} \tag{5}$$

$l_n = (x_n + x_{n-1})/2$ 是两个样点之间的中心点。我们称变换(4)为非均匀离散傅里叶变换(NDFT),NDFT 不能精确地还原重现傅里叶谱。

设 $X = (x_n)_{N \times 1} = (x_0, x_1, \cdots, x_{N-1})^T$ 为已知空间位置向量,在这些点上有观测数据

$Y = (p(x_n))_{N \times 1} = (p(x_0), p(x_1), \cdots, p(x_{N-1}))^T$ 为已知的非均匀数据

现在考虑采样间隔为 Δk_x,数据带限区间在 $[-M\Delta k_x, M\Delta k_x]$ 总共有 $M_p = 2M+1$ 个样点的一个规则采样傅里叶域,对任一空间位置 x_n,与 NDFT 对应的离散反傅里叶变换由下式给出

$$p(x_n) = \frac{\Delta k_x}{2\pi} \sum_{m=-M}^{M} P(m\Delta k_x) e^{-jm\Delta k_x x_n} \tag{6}$$

且是精确的，Δk_x 的选择应足够小以避免空间 x 上的假频。将不规则空间采样位置 $(x_0, x_1, \cdots, x_{N-1})$ 所对应的方程(6)的 N 个式子结合起来可写成矩阵形式

$$\boldsymbol{Y} = \boldsymbol{AP} \tag{7}$$

式中的 $\boldsymbol{P} = (P(m\Delta k_x))_{M_p \times 1}$，

$$\boldsymbol{A} = \left(\frac{\Delta k_x}{2\pi} \exp(-jm\Delta k_x x_n)\right)_{N \times M_p}, \quad m = -M, \cdots, M, \quad n = 0, \cdots N-1$$

利用(7)式求 $\boldsymbol{P}$ 是一个欠定的线性反演问题，存在解的非唯一性，即对数据空间中已知的观测结果 $\boldsymbol{Y}$，在模型空间中存在多个可能的 $\boldsymbol{P}$，它们都能表示 $\boldsymbol{Y}$。按反演理论，方程(7)一般由含有正则化项的目标函数求解，构造目标函数

$$J(\boldsymbol{P}) = S(\boldsymbol{P}) + \frac{1}{2\sigma_n^2} \|\boldsymbol{Y} - \boldsymbol{AP}\|_2^2 \tag{8}$$

式中 σ_n^2 是观测数据中所含噪声的方差，这里设噪声服从高斯分布 $N(0, \sigma_n^2)$。$S(\boldsymbol{P})$ 是用柯西(Cauchy)分布施加的正则化项，

$$S(\boldsymbol{P}) = \sum_{k=1}^{M_p} \ln\left(1 + \frac{P_k P_k^*}{2\sigma_P^2}\right) \tag{9}$$

它是功率向量 $PV_k = P_k P_k^*, k = 1, 2, \cdots, M_p$ 稀疏性的一种度量。反演能够达到的稀疏量由常量 σ_P 控制，也取决于噪声级别。将(9)代入(8)得

$$J(\boldsymbol{P}) = \sum_{k=1}^{M_p} \ln\left(1 + \frac{P_k P_k^*}{2\sigma_P^2}\right) + \frac{1}{2\sigma_n^2} \|\boldsymbol{Y} - \boldsymbol{AP}\|_2^2 \tag{10}$$

目标函数最小化可获得正则化的唯一解 $\hat{\boldsymbol{P}}$，对 $J(\boldsymbol{P})$ 求导并令导数等于零(见附录)，给出下面的估计式

$$\hat{\boldsymbol{P}} = (\boldsymbol{A}^H \boldsymbol{A} + \lambda \boldsymbol{Q}^{-1})^{-1} \boldsymbol{A}^H \boldsymbol{Y} \tag{11}$$

式中 $\lambda = \frac{\sigma_n^2}{\sigma_P^2}$，$\boldsymbol{Q}$ 是一个 $M_p \times M_p$ 的对角矩阵，对角元素

$$q_{ll} = 1 + \frac{P_l P_l^*}{2\sigma_P^2}, \quad l = 1, 2, \cdots, M_p \tag{12}$$

方程(11)类似于阻尼最小平方解，但对角阵 $\boldsymbol{Q}$ 的元素非线性地与傅里叶系数 P_k 相联系，因此，需要迭代计算估计量 $\hat{\boldsymbol{P}}$。

使用恒等式

$$\boldsymbol{A}^H(\lambda \boldsymbol{I}_N + \boldsymbol{A}\boldsymbol{Q}\boldsymbol{A}^H) = (\boldsymbol{A}^H \boldsymbol{A} + \lambda \boldsymbol{Q}^{-1})\boldsymbol{Q}\boldsymbol{A}^H \tag{13}$$

$$(A^H A + \lambda Q^{-1})^{-1} A^H = Q A^H (\lambda I_N + A Q A^H)^{-1} \tag{14}$$

我们可把(11)式重写为

$$\hat{P} = Q A^H (A Q A^H + \lambda I_N)^{-1} Y \tag{15}$$

由于 Q 依赖于 $\hat{P}$，上面的方程必须利用迭代算法求解。

2　模型数据的迭代非线性估计

方程(11)需要一个 $M_p \times M_p$ 矩阵的反演，而方程(15)需完成一个 $N \times N$ 矩阵的反演，迭代求解(15)式的计算优势是显而易见的。

首先将(15)式重写为

$$\hat{P} = Q A^H (A Q A^H + \lambda I_N)^{-1} Y = Q A^H b$$

式中辅助向量 $b \in R^N$ 由方程组

$$(A Q A^H + \lambda I_N) b = Y \tag{16}$$

的解得到。迭代算法从观测数据的 DFT $P^{(0)}$ 开始，该初始值也用于产生矩阵 $Q^{(0)}$。在每次迭代中，我们计算

$$b^{(\mu-1)} = [A Q^{(\mu-1)} A^H + \lambda I_N]^{-1} Y \tag{17}$$

接着用它更新 DFT

$$P^{(\mu)} = Q^{(\mu-1)} A^H b^{(\mu-1)} \tag{18}$$

式中 μ 表示迭代次数。当

$$\frac{|J_{cg}^{(\mu)} - J_{cg}^{(\mu-1)}|}{[|J_{cg}^{(\mu)}| + |J_{cg}^{(\mu-1)}|]/2} \leqslant \varepsilon （容许偏差）$$

时，停止迭代，取 $\hat{P} = P^{(\mu)}$。一般经过为数不多的几次迭代（≤10）即可达到要求。

一旦获得了尖锐的 DFT 谱，我们就可以对它实现反傅里叶变换产生数据空间中期望的规则采样信号输出。

以上给出的是空间傅里叶变换谱估计的原理，时间采样信号傅里叶变换的数学过程是类似的，只不过指数所用符号的约定不同，此处不在叙述。按照给出的算法编写的处理程序既适用于空间采样信号，也适用于时间采样信号，在算法的实现程序中设置了一个参数 xt，当 $xt = 1$ 时处理的是空间采样信号，而当 $xt = -1$ 时处理的是时间采样信号。

3 数值计算试验
3.1 合成数据试验

图 1 画出了一个简单的不规则采样正弦信号,采样点的位置由随机函数

$$x_l = 128 * \Re \quad l = 1,2,\cdots,128 \tag{19}$$

生成,式中 $\Re$ 是随机数,用 NRNDS.FOR 产生[8]。将 x_l 按升序重排序,用函数

$$p(x_l) = \sin(\frac{x_l}{5}) \tag{20}$$

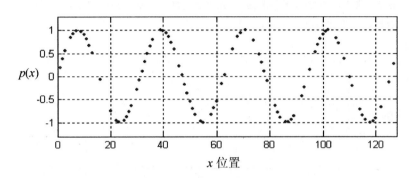

图 1 一个不规则采样正弦函数

产生合成数据。这个正弦波在横坐标轴(0,128)上持续大约 4 个周期,是一个在高度不规则网格上的低频信号。利用(4)式计算的傅里叶谱显示在图 2 中,实际的谱是处处非零的,扩散延伸在傅里叶求和区间中几乎所有的频率上,这种现象就是人们熟悉的频谱泄漏,能量的泄漏是由采样的不规则性和边界效应所造成的,正是数据中信号(脉冲谱)的能量在不规则网格上进行求和时泄漏到所有其它的频率上。漏谱的振幅大部分约是信号谱振幅的 4～20%,最大泄漏比信号频谱振幅的 60% 还要多。根据图 2 中给出的谱恢复原始的不规则间隔数据集是可能的,但要想恢复规则网格上简单正弦波的缺失数据是办不到的,因为正确的内插谱应当是傅里叶域中的一个单脉冲。

图 3 显示出了用本文提出的方法计算的图 1 中数据的谱,表现为一个单一频率的脉冲,几乎无泄漏。有了这样的傅里叶系数,人们就可以利用普通的 FFT 算法容易地重建规则网格上的原始信号。

为了试验本文方法对有不同频率和不同振幅数据的有效性,我们利用下面的函数

$$p(x_l) = 5\sin(\frac{x_l}{5}) + 1.2\sin(x_l + 1) \tag{21}$$

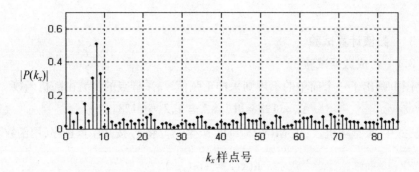

图2 图1中函数的NDFT谱,只画出正的部分。水平轴是波数采样点号,采样点号乘波数采样间隔为波数。

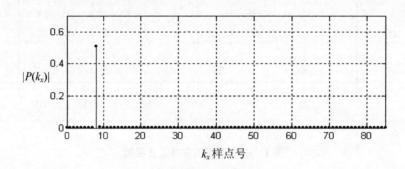

图3 用本文方法得到的谱

抽样位置 x_l 与图1相同,图4显示生成的输入数据,图5显示用NDFT式(4)计算的谱,在这张谱图上只有一个信号成分(最强的信号成分)可以辨明,弱信号的谱隐藏在漏谱中,最大的漏谱振幅达到最强信号的64%,利用这样的傅里叶系数在规则网格上重建是困难的。本文方法的结果显示在图(6),在谱中我们能够看到两个不同的尖峰,这时的傅里叶系数可用于在任意规则网格上的信号重建。

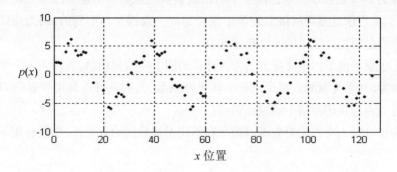

图4 (21)式中函数的不规则采样

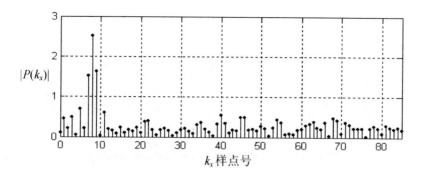

图 5　在不规则网格上 NDFT 的结果仅显示出一个突出的尖峰

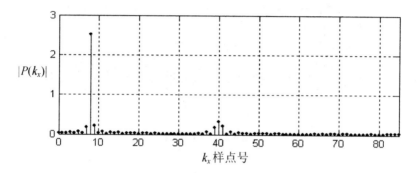

图 6　用本文方法得到的结果,有两个清楚的尖峰

在实际数据中,噪声的存在可能会对任何算法的稳定性造成问题,为了展示本文方法的稳健性,我们考虑下面具有加性随机噪声的函数:

$$p(x_l) = 5\sin(\frac{x_l}{5}) + 1.2\sin(x_l + 1) + 3(\Re_l - 0.5) \tag{22}$$

式中 $\Re_l$ 表示与样点 l 相联系的随机噪声,随机噪声的振幅在$(-1.5,1.5)$的区间上,比第二个信号的振幅更强。使用与前两个试验中相同的不规则抽样网格,不规则采样数据显示在图7,与图4中显示的数据相比,图7中的数据有更多的振荡,且最大的绝对值大约是6.5。

非均匀离散傅里叶变换(NDFT)的谱可利用(4)式得到(见图8),输入数据中加入了随机噪声,但最强的信号谱与图5相比没有多大变化,这一事实可能的解释是噪声是随机的。图9是用本文方法得到的结果,它表明,该方法在有随机噪声时是稳定的,信号的两个可辨别的峰值在谱中是清楚可见的。

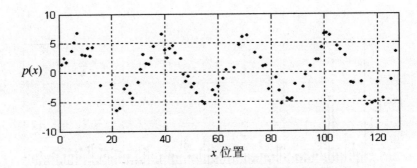

图 7　两个正弦函数与随机噪声的不规则采样叠加

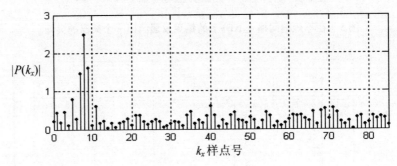

图 8　NDFT 的谱

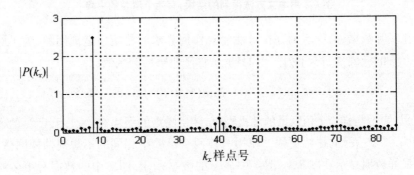

图 9　用本文方法得到的谱

3.2　应用于缺失数据重建

非均匀采样信号的谱估计和缺失数据的重建是时间序列分析中的两个重要问题,在这里通过一个实际例子说明提出的算法可用来解决这两个问题。

图 10 绘出了一个由 400 个样值组成的时间序列,其中有两个频率分别为 0.0795 和 0.0954 的单位振幅谐波,时间序列被 $\sigma_n = 0.2$ 的高斯噪声所污染。从时

间序列中提取三段(第一段有 20 个点,第二段和第三段有 40 个点)产生一个有缺失数据的序列(见图 11),以它作为已知的不规则采样序列,让我们用本文的方法来恢复重建未采样空白区的缺失数据。用普通的 NDFT 求和式所得到的功率谱如图 12 所示,采用文中提出的算法估计的功率谱如图 13 所示(10 次迭代),由它重建得到的时间序列显示在图 14 中,重建的时间序列中占总长度 75% 的空缺数据得到很好的恢复。图 15 显示的是原始的合成时间序列与重建结果的差,这也是合成时间序列中所用噪声序列的估计。

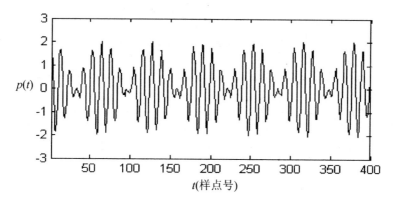

图 10　一个时间序列

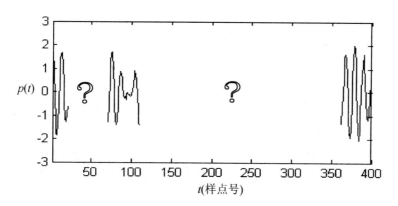

图 11　有空缺数据的部分时间序列采样值

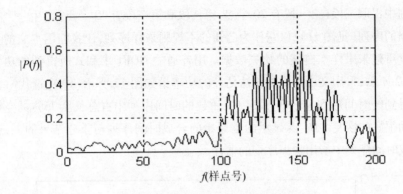

图12　图11中样点值的NDFT谱

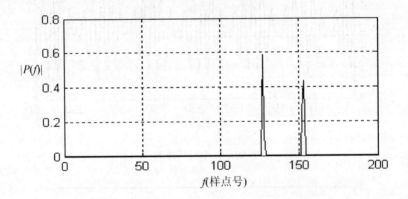

图13　用本文方法计算的图11中样点值的谱

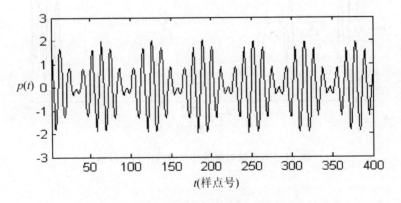

图14　由图13的谱重建恢复的时间序列

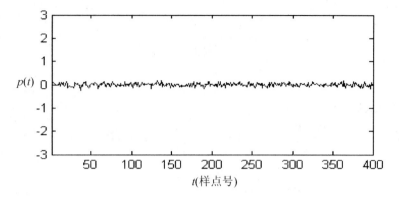

图 15　图 14 与图 10 的差

4　结束语

不规则采样或缺失数据的傅里叶重建是信号正则化处理的主要方法之一,其关键是精确估计信号的傅氏谱,面临的主要任务是如何压制频谱能量的泄漏。文中提出一种能抑制非均匀离散傅里叶变换频谱泄漏的方法,提高了 DFT 的分辨率。实际计算试验结果显示了这种方法的有效性,利用该方法能较好地重建在时间(或空间)方向上缺失的数据。

参考文献

[1] 程佩青. 数字信号处理[M]. 第三版. 北京:清华大学出版社,2007:98 - 195.

[2] Oppenheim A V, Schafer R W, Buck J R. Discrete - time Signal Processing[M]. Second edition. Prentice - Hall,1999.

[3] 陈明凯,郑翔骥,汪晓强. 减少频谱泄漏的一种新的等角度间隔采样递推算法[J]. 电工技术学报,2005,20(8):94 - 98.

[4] 王晓君,陈禾,仲顺安. 改进的脉冲雷达信号加窗 DFT 频谱校正算法[J]. 北京理工大学学报,2008,28(2):164 - 167.

[5] Smith D C, Nelson D J. Detection and resolution of narrow band signal components by concentrating the DFT[C]//5th International Workshop on Information Optics(WIO'06). AIP Conference Proceedings, Volume 860. Toledo, Spain: WIO,2006:200 - 209.

[6] 高清维,程蒲,张道信. 基于对称延拓的 DFT 频谱泄漏抑制方法[J]. 安徽大学学报(自然科学版),2000,24(2):65 - 67.

[7] Lyons R G. 数字信号处理[M]. 第二版. 朱光明,程建远,刘保童,等译. 北京:机械工业出版社,2006:46 - 56.

[8] 徐士良. FORTRAN 常用算法程序集[M]. 北京:清华大学出版社,1995:502 - 504.

附录

(11)式的证明

$$\frac{\partial J(\boldsymbol{P})}{\partial \boldsymbol{P}^H} = \frac{\partial}{\partial \boldsymbol{P}^H}\Big[S(\boldsymbol{P}) + \frac{1}{2\sigma_n^2}\parallel \boldsymbol{Y}-\boldsymbol{AP}\parallel_2^2\Big]$$

$$= \frac{\partial S(\boldsymbol{P})}{\partial \boldsymbol{P}^H} - \frac{1}{2\sigma_n^2}\boldsymbol{A}^H(\boldsymbol{Y}-\boldsymbol{AP}) \tag{A1}$$

为计算正则项 $S(\boldsymbol{P})$ 的导数，简便起见，我们计算关于向量 $\boldsymbol{P}^H$ 的元素的导数

$$\frac{\partial S(\boldsymbol{P})}{\partial P_k^*} = \frac{\partial}{\partial P_k^*}\sum_{k=1}^{M_p}\ln\Big(1+\frac{P_k P_k^*}{2\sigma_P^2}\Big)$$

$$= \frac{1}{2\sigma_P^2}\Big(1+\frac{P_k P_k^*}{2\sigma_P^2}\Big)^{-1}P_k$$

$$= \frac{1}{2\sigma_P^2}q_{kk}^{-1}P_k \tag{A2}$$

(A2)写成矩阵形式,有

$$\frac{\partial S(\boldsymbol{P})}{\partial \boldsymbol{P}^H} = \frac{1}{2\sigma_P^2}\boldsymbol{Q}^{-1}\boldsymbol{P} \tag{A3}$$

把(A3)代入(A1)并令其等于零,我们得到

$$\hat{\boldsymbol{P}} = (\boldsymbol{A}^H\boldsymbol{A} + \lambda\,\boldsymbol{Q}^{-1})^{-1}\boldsymbol{A}^H\boldsymbol{Y} \tag{A4}$$

注：本文曾发表在2010年2月《西北师范大学学报》第2期上

一种随机振动信号幅度、相位及零基线漂移分离方法

张少刚　李芙蓉　赵小龙[*]

为准确得到随机振动信号各分量幅度、相位及零基线漂移,通过 Hilbert 变换,随机信号被映射成复平面上的曲线,该曲线看成由随时间可变长度矢量末端绕动点逆时针转动所形成,其中动点做随机运动。由于各分量具有不同频率,反映在复平面具有不同的转动角速度,因此各分量之间具有相对运动,其中低频分量不改变高频分量的曲率,曲率是随机振动信号各分量的不变量,利用该性质,将曲率确定的信号分量逐步分离,得到单分量随机振动信号,残留部分为振动信号零基线漂移,从而自适应求解出随机振动信号随机幅度、相位和零基线漂移量成分。曲线的密切圆半径就是该曲线曲率的倒数,算法上利用密切圆的拟合得到的曲率半径替代计算曲率的求导运算,减少数据离散性带来的误差对计算结果的影响。单分量信号幅度、相位变化规律是进一步研究复杂随机信号基础、也为工程应用提供一种可借鉴的计算方法。

1 引言

随机振动信号分析处理一般采用傅里叶变换或者统计特征量研究其变化规律。对于非平稳随机振动信号,利用平稳化、局部平稳化或者借助非线性随机信号分析的方法进行研究,典型的有 Wigner – Ville 分布、谱分析、小波分析、盲源分离和高阶统计量分析,上述方法建立在主观设置一种特定的变换核函数的基础上[1-2],虽然能解决一些特定问题,但整体仍有很多问题需要深入研究。经验模态分解(Empirical Mode Decomposition,简称 EMD)是由黄锷(N. E. Huang)与他人于 1998 年创造性地提出的一种新型自适应信号时频处理方法,EMD 方法被认为

---

[*] 作者简介:张少刚(1961—),男,辽宁海城人,天水师范学院教授、硕士。主要从事信息处理的研究。

是 2000 年来以傅立叶变换为基础的线性和稳态频谱分析的一个重大突破,该方法是依据数据自身的时间尺度特征来进行信号分解,无须预先设定任何基函数。这一点与建立在先验性的谐波基函数和小波基函数上的傅里叶分解与小波分解方法具有本质性的差别。正是由于这样的特点,EMD 方法在理论上可以应用于任何类型的信号的分解,因而在处理非平稳及非线性数据上,具有非常明显的优势,适合于分析非线性、非平稳信号序列,具有很高的信噪比。但该方法没有给出分解理论依据[3]。本文在此基础上给出随机振动信号在复平面的分解方法,也是在理论上验证 Hilbert – Huang 变换的正确性。多分量振动信号分量之间互相关系以及与零基线漂移的分离是信号深入信号结构特征的基础。设随机振动信号 $s(t)$ 可表示为

$$s(t) = \sum_{i=1}^{n} a_i(t) cos\varphi_i(t) + r(t) \tag{1}$$

其中 $a_i(t)cos\varphi_i(t)$ 表示 $s(t)$ 的一个分量。$a_i(t) > 0$、$\varphi_i(t)$、$r(t)$ 分别表示幅度、相位、零基线漂移量的随机变量,零基线漂移量表示交变成分平衡点的随机变化,这种表达不具有唯一性。对随机振动信号 $s(t)$ 进行 Hilbert 变换得到 $H[s(t)]$,$s(t)$ 与 $H[s(t)]$ 构成一变换对,令 $z(t) = s(t) + iH[s(t)]$,$z(t)$ 唯一确定随机振动信号 $s(t)$[4]。在一定时间长度内,分离各分量以及漂移量,对每个分量的幅度和相位进行分离,对进一步揭示随机振动信号内在规律,为随机振动信号深入研究应用提供可以一种可借鉴的方法。

2 理论分析

如果将连续随机振动信号 s(t) 看成是非线性时变物理可实现系统的输出结果,通过希尔伯特变换可构造

$$z(t) = s(t) + iH[s(t) \tag{2}$$

$z(t)$ 具有唯一的表达形式,$z(t)$ 可理解为随时间可变长度矢量末端绕动点逆时针转动形成的轨迹曲线,$s(t)$ 是 $z(t)$ 横坐标轴的投影,动点做随机运动,在物理可实现系统的前提下,其凹向始终在 $z(t)$ 左侧。换言之,$z(t)$ 是 $s(t)$ 复平面的唯一表达,这两者的推论完全基于 $s(t)$ 是物理可实现系统和 Hilbert 变换,两者等效[4-5]。将式(1)代入式(2)并整理

$$z(t) = \sum_{i=1}^{n} [a_i(t)cos\varphi_i(t) + i a_i(t)sin\varphi_i(t)] + r(t) \tag{3}$$

其中随机时变幅度 $a(t)$ 表示随机振动信号交变部分能量变化,是标量函数,也可理解为复平面随时间可变长度矢量绕动点 $r(t) + iH[r(t)]$ 运动的轨道半

径[6]。$\varphi(t)$ 表示单分量随机振动信号交变成分相位变化,对应复平面质点绕动点的转动角。$\varphi(t)$ 的导数表示交变部分的瞬时频率。$r(t)$ 表示随机振动信号零基线漂移量,$r(t)+iH[r(t)]$ 表示复平面动点的平面随机运动。由于系统具有非线性随机时变特性,逆时针转动可变长度矢量末端形成的轨迹任意时刻,可以理解质点以曲率半径逆时针转动与动点随机运动合成,由于动点的运动是相对较低频率成分产生的,故动点运动只是引起相对高频成分的矢量的整体平动,动点的运动不改变其相对高频成分形成曲线的曲率,质点定向平面运动可以用曲率半径随机变化、圆心做随机运动、一系列前后相依的瞬态曲率密切圆表示,随机振动信号就是上述质点运动在坐标轴的投影。这样可以把时间域随机振动信号问题转换成复平面动态矢量运动几何问题[7]。基于上述理解,动态矢量运动形成的曲线处处具有不同的曲率,该曲率唯一确定了质点绕动点一系列瞬时转动,根据运动合成原理与微分几何曲率的定义[8],动点的随机运动只改变曲线运动轨迹的方位,不影响曲线轨迹的曲率,曲率由每个时刻的角速度最大的分量决定。已知总体轨迹 $z(t)$,逐点求解 $z(t)$ 的曲率,可以得到每个时刻瞬时曲率半径 $a_i(t)$ 和该瞬时曲率半径转动角 $\varphi_i(t)$,该系列曲率特性确定第一分量,剩余部份就是动点的随机运动成份,随机运动成份实部表达 $s(t)$ 随机漂移量[9],对于多分量振动信号可以重复上述过程得到第二分量,以此类推,求解曲率的过程同时也完成 $a_i(t)$、$\varphi_i(t)$、$r(t)$ 分三个部分的分离,在一个有限长度内确定的随机振动信号观测样本进行随机变量的分离。对这些随机变量的独立变化以及之间的相关性研究,可以进一步探索随机振动信号局部特征和整体变化规律。

3 算法描述与模拟算例
3.1 算法描述
由于实际随机振动信号都是通过采样得到的离散的观测值,曲率的求解涉及微分运算,对于离散序列,用差分代替微分运算会带来一定的误差,拟采用局部密切曲率圆拟合算法获得曲率半径,曲率半径的倒数即为曲率[10-12]。算法步骤如下:

①获取离散随机振动信号序列 $s(t_i)$,$i(i=1,2,\cdots n;)$,i 是序列下标;

②对 $s(t_i)$ 进行 Hilbert 变换得到 $H[s(t_i)]$,构造 $z(t_i)=s(t)+iH[s(t)]$;

③对 $z(t_i)$(至少三个连续点)逐点拟合密切曲率圆,进而得到瞬时曲率半径序列 $a(t_i)$,也是随机振动的随机幅度序列;

④逐点拟合曲率半径 $a(t_i)$ 的瞬时转动角 $\varphi(t_i)$,也是随机振动信号随机相位,得到随机振动信号瞬时相位序列,根据单分量信号定义,$a(t_i)cos\,\varphi(t_i)$ 为原随

机振动信号 $s(t_i)$ 第一分量信号;

⑤$z(t_i)$ 减去 $a(t_i)$ 成份得到 $r(t_i)$,$r(t_i)$ 实部反映 $s(t_i)$ 的零基线漂移量,如果对于多种模态混合的多分量信号,可以对剩余成分 $r(t_i)$ 重复上述过程,得到其余分量及其参数,直至前后两次分离曲率差值相近,建议经验值小于等于 0.5 时,终止分离,剩余量为单纯漂移量。

3.2 模拟算例

利用 matlab 的 unifrnd(参数)函数产生模拟振动信号序列 $s(t_i)$,$i=1\cdots n$,n 为序列数据数量,参数分别取:时间长度 10 秒;时间间隔 0.07 秒,随机幅值范围 0-8(无量纲)。见图 1。

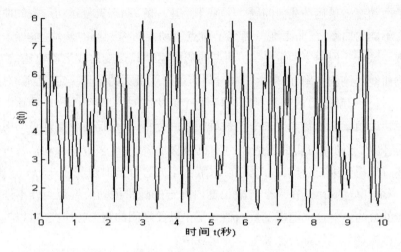

图 1 随机振动信号 $s(t_i)$

利用 matlab 下的 $Hilbert(s(t_i))$ 得到对随机振动信号的希尔伯特变换,构建 $z(t_i) = s(t_i) + Hilbert(s(t_i))$,$z(t_i)$、$z(t_{i+1})$、$z(t_{i+2})$($i=1\cdots n-2$) 逐点拟合圆,拟合方法可以采用平均法或者最小二乘法,每次拟合的曲率半径即为 $a(t_i)$,即随机振动信号幅度序列。见图 2。通过逐个计算 $a(t_i)$、$a(t_{i+1})$ 的转动角得到 $\varphi(t_i)$ 序列($i=1\cdots n-3$),取其主值。见图 3。

将 $a(t_i)$ 序列与 $\varphi(t_i)$ 点乘得到 $a(t_i)\cos\varphi(t_i)$,其满足单分量信号定义的一个分量。见图 4.

令 $z_1(t_i) = a(t_i)\cos\varphi(t_i) + i\,a(t_i)\sin\varphi(t_i)$,则 $r(t_i) = z(t_i) - z_1(t_i)$,如果 $r(t_i)$ 不包含其他分量振动信号,则其本身就是零基线漂移量,如果 $r(t_i)$ 还包含相对较低成分信号分量,重复上述步骤,直至曲率值小于 0.5,停止分解。本算例只分解一次,量基线漂移量见图 5。

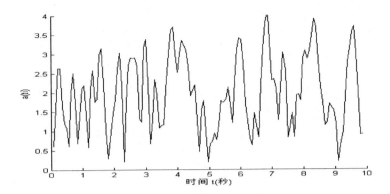

图 2 随机幅度 $a(t_i)$

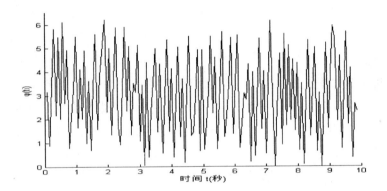

图 3 随机相位主值 $\varphi(t_i)$

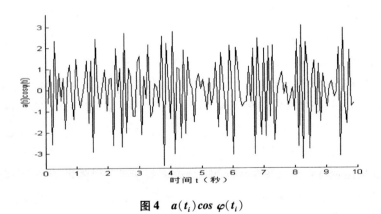

图 4 $a(t_i)\cos\varphi(t_i)$

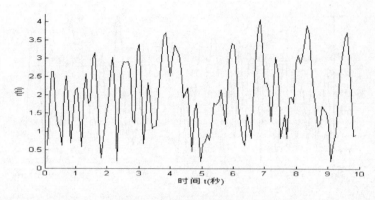

图 5　零基线漂移量 $r(t_i)$

4　结论

随机振动信号经 Hilbert 变换后,映射成复平面运动轨迹,根据运动合成原理与微分几何中曲率的定义,曲线整体平动和转动不改变曲线曲率的性质,通过特征不变量曲率将随机振动信号自适应分解单纯的交变分量与零基线漂移部分,对于多分分量信号,可以反复分离,得到多个分量和零基线漂移部分,对于每个分量可以精确计算出随时间变化的随机幅度、相位和零基线漂移量三部分,每个分量有明显的直观物理意义,在特定应用背景下具有可解释性,随机幅度代表交变部分能量的变化,随机相位的导数代表瞬时频率,表达振动瞬时变化激烈程度,零基线漂移部分代表各种因素导致系统平衡态的变化。初步研究结论表明,零基线漂移量对随机振动信号幅度、相位的变化有不同程度的影响,尤其是对瞬时频率的计算影响较大,对随机振动信号剔除零基线漂移部分可以进一步得到准确的时频表达,不会出现无法解释的负频率成分,按有意义的物理变量进行分解,分解作为统计分析前的预处理,不失为一种新的探索。本文仅仅是一种分离方法的研究,在振动信号分析、机械故障诊断等应用背景下,为随机振动信号分析、处理提供参考借鉴。进一步研究随机幅度、相位和零基线漂移量的统计规律,以及它们之间的相关性,可能获得更有意义的结果。随机幅度与随机相位协同变化规律以及随机幅度变化是否具有离散特性有待深入研究。对于多分量随机信号,一次分离后的残余量还含有低频分量,用上述方法反复分离,得到各个分量,其分离终止条件取决于两次分离的曲率差值,其更深层次规律需要进一步研究。

参考文献

[1] 李舜酩,郭海东,李殿荣.振动信号处理方法综述[J].仪器仪表学报,2013,34(8):1907-1915.

Li Shunming,Guo Haidong,Li Dianrong. Review of vibration signal processing methods[J]. Chinese Journal of Scientific Instrument,2013,34(8):1907-1915.

[2] 丁康,陈健林,苏向荣.平稳和非平稳信号的若干处理方法及发展[J].振动工程学报,2003,16(1):1-10.

Ding Kang,Chen Jianlin,Su Xiangrong. Development in Vibration Signal Analysisand Processing Methods[J]. Journal of Vibration Engineering,2003,16(1):1-10.

[3] 杨永锋,吴亚锋,经验模态分解在振动分析中的应用[M].北京:国防工业出版社,2013:5-24.

Yang Yongfeng,Wu Yafeng. Applications of Empirical Mode Decomposition in Vibration Analysis[M]. Beijing:National Defend Industry Press,2013:5-24.

[4] 高新波,刘聪锋,宋骊平,牛振兴.随机信号分析[M].北京:科学出版社,2009:134-170.

Gao Xinbo,Liu Comghui,Song Liping,Niu Zhenxing. Random signal analysis[M]. Beijing:Science press,2009:134-170.

[5] 王宏禹,邱天爽,陈喆.非平稳随机信号分析与处理(第2版)[M].北京:国防工业出版社,2008:18-43.

Wang Hongyu,Qiu Tianshuang,Chen Zhe. Nonstationary Random Signal Analysis and Processing(Second Edition)[M]. Beijing:National Defend Industry Press,2008:18-43.

[6] 胡海峰,胡茑庆,秦国军.基于改进经验AM-FM解调的复杂信号瞬时特征分析方法[J].国防科技大学学报,2011,33(02):119-124.

HU Hai-feng,HU Niu-qing,QIN Guo-jun. Instantaneous Characteristics Analysis Method for Complicated[J]. Journal of National University of Defense Technology,Signals Based on Improved Empirical AM-FM Demodulation,2011,33(02):119-124.

[7] 李振兴,玄志武,徐洪洲.振动信号阶跃零点漂移分析与去除方法研究[J].强度与环境.2008,35(04):61-64

LI Zhen-xing,XUAN Zhi-wu,XU Hong-zhou. Analysis and study on method of wiping off zero excursion for vibration signal[J]. Structure & Environment Engineering. 2008, 35(04):61-64.

[8] 陈平,李庆民,赵彤.瞬时频率估计算法研究进展综述[J].电测与仪表,2006(07):1-7.

CHEN Ping,LI Qing-min,ZHAO Tong. Advances and Trends Instantaneous Frequency Estimation Methodology[J]. Electrical Measurement & Instrumentation,2006(07):1-7.

[9]刘永本.非平稳信号分析导论[M].北京:国防工业出版社,2006:13-32.
Liu Yongben. An Introduction To Nonstationary Signal Analysis[M]. Beijing:National Defend Industry Press,2006:13-32.

[10]孟道骥,梁科.微分几何(第二版)[M].北京:科学出版社,2008:6-34.
Meng Daoji,Liang Ke. differential geometry(Second Edition)[M]. Beijing:Science Press,2008:6-34.

[11]李海林,梁叶,王少春.时间序列数据挖掘中的动态时间弯曲研究综述[J].控制与决策,2018,33(08):1345-1353.
LI Hai-lin,Liang Ye,Wang Shao-chun. A review on dynamic time warping in time series data mining[J/OL]. Control and Decision,2018,33(08):1345-1353.

[12]崔凤午.空间曲线曲率中心轨迹的曲率与挠率[J].武汉科技学院学报,2010,23(02):41-43.
CUI Feng-wu. Curvature and Torsion of Trajectory in Space Curve and Curvature Center[J]. Journal of Wuhan Textile University,2010,23(02):41-43.

注:本文曾发表在2019年10月《应用力学学报》第5期上

Numerical study of the influence of applied voltage on the current balance factor of single layer organic light-emitting diodes

Lu Feiping　Liu Xiaobin　Xing Yongzhong[*]

摘　要:电流平衡因子是器件复合电流和器件总电流之比,是影响器件外量子效率的重要因素,在磷光和荧光有机发光器件的性能中发挥着重要作用。在本论文中,借助单层有机电致发光器件的数值模型,研究了器件的外加偏压对器件电流平衡因子的影响。结果表明,当电子注入势垒和空穴注入势垒相等时,在低电压区,器件的电流平衡因子最大。在高电压区,当电子迁移率大于空穴迁移率时,且电子注入势垒大于空穴注入势垒时,器件的电流平衡因子最大,而当电子迁移率小于空穴迁移率时,情况恰恰相反。当电子迁移率和空穴迁移率相等时,电子注入势垒和空穴注入势垒相等时,器件具有最大的电流平衡因子。除此之外,随着外加偏压的增加,器件的电流平衡因子也增加。本论文获得的结果可为深入了解有机电致发光器件的工作机理及制备高性能器件提供帮助。

Current balance factor(CBF) value, the ratio of the recombination current density and the total current density of a device, is a very important prerequisite when discussing the effect of the other parameters that affect the external quantum efficiency, and plays an important role in fluorescence – based organic light – emitting devices (OLEDs), as well as in the performance of the organic electrophosphorescent devices. In this letter, the influence of theapplied voltage on the CBF value of the single layer OLEDs was investigated by utilizing a numerical model of a bipolar single layer OLED with organic layer trap free and without doping. Results show that the largest CBF value

---

[*] 作者简介:路飞平(1980—　),男,甘肃岷县人,天水师范学院副教授、博士,主要从事有机光电子材料及器件物理研究。

can be achieved when the electron injection barrier(φ_n) is equal to the hole injection barrier(φ_p) in the lower voltage region at any instance. The largest CBF in the higher voltage region can be achieved in the case of $\varphi_n > \varphi_p$ under the condition of electron mobility(μ_{0n}) > hole mobility(μ_{0p}), whereas the result for the case of $\mu_{0n} < \mu_{0p}$, is the opposite. The largest CBF when $\mu_{0n} = \mu_{0p}$, can be achieved in the case of $\varphi_n = \varphi_p$ in the entire region of the applied voltage. In addition, the CBF value of the device increases with increasing applied voltage. The results obtained in this paper can present an in-depth understanding of the OLED working mechanism and help in the future fabrication of high efficiency OLEDs.

1 INTRODUCTION

Organic light-emitting diodes(OLEDs) are widely used in flat-panel displays and solid state lighting because of their unique properties.[1-4] OLEDs must work under a relatively high luminance and must have a long working lifetime to meet practical use. Experiments have demonstrated that the working lifetime of OLED decreases with increasing current density.[5] However, the luminance of OLEDs is proportional to the current density of the devices, which leads to a conflict between the high luminance and the long working lifetime required for OLEDs. Therefore, the luminance must be improved as much as possible under a relatively low current density to solve such problem. One of most effective methods is improving the current balance factor(CBF), which is the ratio of the recombination current density and the total current density of the device. The CBF of OLED is an important prerequisite when discussing the effect of the other parameters affecting external quantum efficiency. A high CBF value guarantees that an OLED will work with high luminance under relatively low current density, so OLEDs can obtain long working lifetime. Researchers always modify the interface of the electrode with a thin layer, such as inserting MnO,[6] LiF,[7] Al_2O_3,[8] or $CsCO_3$[9] into the interface of the cathode/organic layer, or by inserting $CuPc$[10] or MoO_3[11] into the interface of the anode/organic layer, to adjust the injection ability of carriers and improve the CBF of the devices. Single layer OLED is one of the current focuses of research because of their simple fabrication procedures. A previous study used Tris(8-hydroxyquinoline)aluminum(Alq_3) as emission layer to fabricate single layer OLEDs. The results demonstrated that utilizing different cathode materials can change the injection ability of the electrons, thereby changing the CBF of the devices. In addition, the external quantum efficiency of devices increases with increasing applied voltage.[12] A very low turn-on voltage, along with significantly improved luminance and current density of the device can be achieved by depositing an ultra-thin layer of nickel oxide(NiO) on the indium-tin oxide(ITO) anode to enhance hole injection in OLED and improve the CBF value of the devices.[13] A single-layer dendrimer OLED formed by spin-coating exhibited a very high-efficiency green phosphorescence, which can be attributed to the excellent film forming properties and high photoluminescence quantum yield of the dendrimer blend and the efficient injection of charge into the emissive layer.[14] The current paper investigated the influence of applied voltage on the CBF characteristic of single layer OLEDs based on a numerical model of a bipolar single layer OLED with organic layer trap free and without doping. The results obtained can present an in-depth understanding of the working mechanism of OLED and help in future fabrication of high efficiency OLEDs.

2 MODEL

Single layer OLED is a simple device of an organic thin film layer sandwiched between two metallic contacts. One contact is always a metallic cathode and the other is a semitransparent metal-oxide anode for bottom OLEDs, contrarily for top OLEDs. In case of the hole injection at the left($x=0$) and the electron injection at the right($x=L$), the time-independent current continuity equations for the electron and the hole are as follows:

$$\frac{dJ_p}{dx} = q(G-R) \tag{1}$$

$$\frac{dJ_n}{dx} = -q(G-R) \tag{2}$$

where J_p and J_n are the hole and electron current density, respectively; q is the elementary charge; and G and R are the carrier generation and recombination rates. The generation rate is very small and can be ignored because of the high energy band for organic materials. Assuming that the recombination is a Langevin type,[15] the following equation can be obtained:

$$R = \frac{4\pi q}{\varepsilon_0 \varepsilon_r}(\mu_n + \mu_p)(np - n_i^2) . \tag{3}$$

where ε_0 and ε_r are the permittivity of vacuum and relative dielectric constant of the organic materials, respectively; n and p are the electron and hole concentrations, respectively; and μ_n and μ_p are the electron and hole mobilities, respectively, which are always filed-dependent. The following equations can be obtained based on the universal Poole-Frenkel form:

$$\mu_n(E) = \mu_{0n}\exp(\sqrt{\frac{F}{F_{0n}}}) \tag{4}$$

$$\mu_p(E) = \mu_{0p}\exp(\sqrt{\frac{F}{F_{0p}}}) \tag{5}$$

where μ_{0n} and μ_{0p} are the zero field mobilities; F_{0n} and F_{0p} are the characteristic electric fields for electrons and holes, respectively; and n_i is the intrinsic carrier concentration in the organic semiconductor defined by

$$n_i^2 = N_{HOMO}N_{LUMO}\exp(-\frac{E_g}{k_B T}). \tag{6}$$

where N_{HOMO} and N_{LUMO} are the total densities of states of the highest occupied molecular orbitals (HOMOs) and the lowest unoccupied molecular orbitals (LUMOs), respectively; E_g is the energy gap between LUMO and HOMO; k_B is the Boltzmann constant; and T is the working temperature. In the bulk of the organic layer, carriers are transported from one side to another in two ways: one is drift-driven by the applied electric field, whereas the other is through diffusion, a transport from the high carrier concentration region to the low carrier concentration region. Hence, the electron and hole current density in organic bulk can be expressed as follows:

$$J_p = qp\mu_p F - qD_p\frac{dp}{dx} \tag{7}$$

$$J_n = qn\mu_n F + qD_n\frac{dn}{dx} \tag{8}$$

where D_p and D_n are the diffusion coefficients of the holes and the electrons, respectively, and F is the electric field intensity. The total current density of the device can be expressed as follows:

$$J = J_n + J_p. \tag{9}$$

Assuming that the Einstein relation is still valid, the following equation can be obtained:

$$\frac{D_n(F)}{\mu_n(F)} = \frac{D_p(F)}{\mu_p(F)} = \frac{k_B T}{q}. \tag{10}$$

The electric field F and electrostatic potential V are coupled with n and p through the Poisson equation as follows:

$$\frac{d^2V}{dx^2} = -\frac{dF}{dx} = -\frac{q}{4\pi\varepsilon_0\varepsilon_r}(p-n). \tag{11}$$

In a real device, the electrode/organic interfaces are Schottky contacts, and the carrier current density consist of three parts, namely, tunneling current, thermionic emission current, and a back flowing interface recombination current, which is the time reversed process of thermionic emission.[16,17] Specifically, the hole current injection at anode/organic interface ($x=0$) and the electron current injection at cathode/organic ($x=L$) are considered as follows:

$$J_p(0) = J_{pth} + J_{ptu} - J_{pir} \tag{12}$$
$$J_n(L) = J_{nth} + J_{ntu} - J_{nir} \tag{13}$$

where J_{nth} and J_{pth} are the thermionic emission current, J_{nir} and J_{pir} are the interface recombination current, and J_{ntu} and J_{ptu} are the tunneling current for the electron and the hole, respectively. The thermionic emission current can be expressed as follows:[18]

$$J_{th} = AT^2\exp(-\frac{\varphi_B}{kT}) \tag{14}$$

where A is the Richardson's constant and ϕ_B is the interfacial energy barrier. ϕ_B depends on the electric field at the interface because of the image force:

$$\varphi_B = \varphi - q\sqrt{\frac{q|F_{inter}|}{4\pi\varepsilon}}. \tag{15}$$

For the anode/organic interface, φ is the difference of the Femi level of the anode and the HOMO of the organic layer. The cathode/organic interface is the difference of the Femi level of cathode and the LUMO of organic layer. F_{iner} is the interface electric field; $F(x=0)$ for the anode/organic interface; and $F(L)$ for the cathode/organic interface.

The recombination current at the interface of the anode/organic is proportional to the hole concentration at the interface, which can be expressed as follows:

$$J_{pir} = v_p p(0) \tag{16}$$

whereas,

$$J_{n\ ir} = v_e n(d) \tag{17}$$

at the interface of cathode/organic.

The kinetic coefficient v is determined by the detailed balance between thermionic emission and the interface recombination, which are time-reversed processes of each other:

$$v_p = \frac{AT^2}{p_0} ; v_n = \frac{AT^2}{n_0} \tag{18}$$

where n_0 and p_0 is the density of the states that can be occupied by an electron or a hole.

The tunneling current is expressed as follows:[19]

$$J_{tn} = \left(\frac{3q^2 F(d)}{8\pi h \varphi_{B,n}}\right) \exp\left(-\frac{8\pi \sqrt{2qm_n^* \varphi_{B,n}^3}}{(3hF(d))}\right) \tag{19}$$

$$J_{tp} = \left(\frac{3q^2 F(0)}{8\pi h \varphi_{B,p}}\right) \exp\left(-\frac{8\pi \sqrt{2qm_p^* \varphi_{B,p}^3}}{(3hF(0))}\right) \tag{20}$$

where J_{tn} and J_{tp} are the electron and the hole tunneling currents, respectively; $\varphi_{B,n}$ and $\varphi_{B,p}$ are the interfacial energy barrier for the electrons and the holes; h is Planck's constant; and m_n^* (m_p^*) is the electron(hole) effective mass.

The boundary condition for the Poisson's equation is expressed as follows:

$$V - V_{built-in} = \int_0^d F(x)\,dx \tag{21}$$

where $V_{built-in}$ is the built-in potential of the device, which is the difference of the work function of two electrodes. The following can be obtained according to the principle of current continuity:

$$\frac{dJ}{dx} = 0. \tag{22}$$

The recombination current in an OLED can be expressed as follows:

$$J_r = \int_{x=0}^{L} qR(x)\,dx = J_n(L) - J_n(0) = J_p(0) - J_p(L), \tag{23}$$

and the CBF of devices can be expressed as follows:

$$\eta = \frac{J_r}{J}. \tag{24}$$

The above sets of equations are numerically solved by the Scharfetter – Gummel method.[20,21]

3 RESULTS AND DISCUSSIONS

This section discusses the CBF of the OLEDs in three different conditions according to the relation of the electron and the hole mobility values. The parameters used in the simulation are show in Table 1.

Table 1 Parameters used in the simulation

$N_{LUMO}(cm^{-3})$	1×10^{21}	$N_{HOMO}(cm^{-3})$	1×10^{21}
$F_{p0}(V \cdot cm^{-1})$	5×10^{5}	$F_{n0}(V \cdot cm^{-1})$	5×10^{5}
$n_0(cm^{-3})$	1×10^{21}	$p_0(cm^{-3})$	1×10^{21}
$T(k)$	300	$V(V)$	20
$d(nm)$	100	$E_g(eV)$	3.0
ε_r	3		

3.1 $\mu_{0n} > \mu_{0p}$

Fig. 1 shows the electric field intensity (Fig. 1 - a), the hole and electron concentration (Fig. 1 - b) distribution in the bulk of organic layer, the relationship between applied voltage and CBF value (1 - c), and the relationship between applied voltage and hole and electron injection current density (1 - d) in the cases of electron injection barrier (φ_n) >, = and < hole injection barrier (φ_p) for a single layer OLED with electron mobility (μ_{0n}) larger than the hole mobility (μ_{0p}). As shown in Fig. 1 - c, the largest CBF value can be achieved at the lower applied voltage region when $\varphi_n = \varphi_p$ because the electron and the hole have the nearly same injection ability when $\varphi_n = \varphi_p$ under a lower applied voltage. Under such condition, few carriers can be injected into the organic layer, which transport into the organic film bulk without accumulating at the interface of the electrode/organic. In this case, devices have the largest recombination current and CBF compared with the cases of $\varphi_n > \varphi_p$ and $\varphi_n < \varphi_p$.

More carriers can be injected into the organic thin film with the increase in the applied voltage of the devices. The injected carriers accumulate at the vicinity of the anode/organic interface because of the low mobility of the organic film (as shown in Fig. 1 - b). The number of injected holes are less than the number of injected electrons when $\varphi_n < \varphi_p$, as shown in Fig. 1 - b. Meanwhile, holes accumulate at the vicinity of the anode/organic interface because of their lower mobility compared with the electrons (as shown in Fig. 1 - b). The accumulated holes generate an inner electric field with an opposite direction to the applied electric field, leading to the interface electric field intensity being lower than the initial value (= V/L, where V is the applied voltage and L is the thickness of organic layer). The greater the accumulation at the vicinity of the anode/organic interface, the lower the interface electric field intensity (as shown in Fig. s 1 - b and 1 - a), thereby reducing the injection ability of the holes. Meanwhile, more electrons can be injected into organic layers because the injection barrier is lower; however, the injected electrons can easily transport from the interface into the organic bulk compared with hole because the electron mobility is larger than the hole. Ultimately, the accumulation of electrons at the vicinity of the cathode/organic interface is reduced. In this situation, the carriers injected from both electrodes are deeply imbalanced, leading to the electron injection current ($J_n(L)$) and hole injection current ($J_p(0)$) is deeply imbalanced, as shown in Figs. 1 - b and 1 - d. In this In this case, devices have the smallest CBF compared with the cases of $\varphi_n = \varphi_p$ and $\varphi_n > \varphi_p$.

When φ_n increases and φ_p decreases until $\varphi_n = \varphi_p$, the injection ability of the electron is reduced and the injection ability of the hole is improved. The injected electrons can easily transport from the interface into the organic bulk because $\mu_{0n} > \mu_{0p}$. The hole accumulation at the vicinity of the anode/organic interface is stronger than the electron accumulation at the vicinity of the cathode/organic interface (as shown in Fig. 1 - b), thereby leading to the imbalance of carrier injection from both electrodes. However, compared with the case $\varphi_n < \varphi_p$, the difference between the electrons injection ability and holes injection ability is reduced, so the CBF value is improved compared with $\varphi_n < \varphi_p$.

When φ_n continues to increase and φ_p continues to decrease until $\varphi_n > \varphi_p$, the electron injection ability is further reduced and the hole injection ability is further improved. The accumulation of carrier at the vicinity of the anode and cathode is different because of the different mobilities of the holes and the electrons, resulting in different interface electric field intensities. Nevertheless, a case wherein the carriers injected from both electrodes will be equal may also exist, the difference between the electrons and holes injection ability is further reduced (as shown in Figs. 1 - b and 1 - d), so devices have the largest CBF compared with the cases of $\varphi_n > \varphi_p$ and $\varphi_n < \varphi_p$.

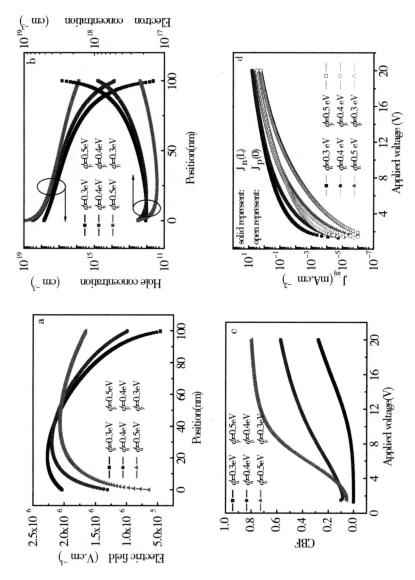

FIG. 1. Calculated distributions of (a) the electric field intensity; (b) hole and electron concentration in bulk of device; (c) the relationship between applied voltage and CBF value; (d) and the relationship between applied voltage and hole and electron injection current density. The parameters used are: $\mu_n = 1 \times 10^{-5} cm^2 \cdot V^{-1} \cdot s^{-1}$, $\mu_p = 1 \times 10^{-6} cm^2 \cdot V^{-1} \cdot s^{-1}$.

Moreover, the CBF value increases in the relative high applied voltage region with increasing applied voltage for all cases ($\varphi_n > \varphi_p$, $\varphi_n = \varphi_p$ and $\varphi_n < \varphi_p$), as shown in Fig. 1 – c. Considering $\varphi_n < \varphi_p$ as an example, the carrier injection ability of an electron from the cathode is more effective than the hole from the anode under a given applied voltage. Given the lower carrier mobility of organic layers, the number of carriers that accumulate at the vicinity of the electrode/organic interface will increase when more carriers are injected into the organic layer, thereby generating a stronger inner electric field. The interface electric field intensity at the cathode/organic becomes weaker (as shown in Fig. 1 – a), thereby reducing the injection ability of electrons. By contrast, the number of accumulated holes at the anode/organic interface is lower. The interface electric field intensity of the anode/organic interface may be larger than the cathode/organic interface. This difference results in the rate of increment of the hole injection current density being larger than the rate of increment of the electron injection current density as the applied voltage increases (as seen shown in Fig. 1 – d). Therefore, the CBF of devices increases with increasing applied voltage. The cases of $\varphi_n = \varphi_p$ and $\varphi_n > \varphi_p$ have similar physical process.

3.2 $\mu_{0n} = \mu_{0p}$

For a single layer OLED with $\mu_{0n} = \mu_{0p}$, Fig. 2 shows the electric field (Fig. 2 – a), the hole and electron concentration (Fig. 2 – b) distribution in the bulk of organic layer, and the relationship between applied voltage and CBF value (2 – c) under the condition that $\varphi_n > \varphi_p$, $\varphi_n = \varphi_p$, and $\varphi_n < \varphi_p$. In the cases of $\varphi_n = \varphi_p$, Fig. 2 – c shows that devices have the largest CBF value compared with when $\varphi_n > \varphi_p$ and $\varphi_n < \varphi_p$, Fig. 2 – a shows that the electric field intensity of the electrode interface for both electrode/organic interfaces is equal when $\varphi_n = \varphi_p$, electrons and holes have the same injection ability and mobility, which allows the carrier concentration to have a symmetrical distribution characteristic (as seen in Fig. 2 – b). Therefore, the device can achieve the largest CBF value. However, when $\varphi_n \neq \varphi_p$, the hole and electron injection ability will be different according to Equations (14), (19) and (20), so the CBF of the device will be decreased compared with the that when $\varphi_n = \varphi_p$. At the same time, the CBF value increases with increasing applied voltage because the carriers injected from both electrodes are equal when the applied voltage increases, similar to that when $\mu_{0n} > \mu_{0p}$, as mentioned previously.

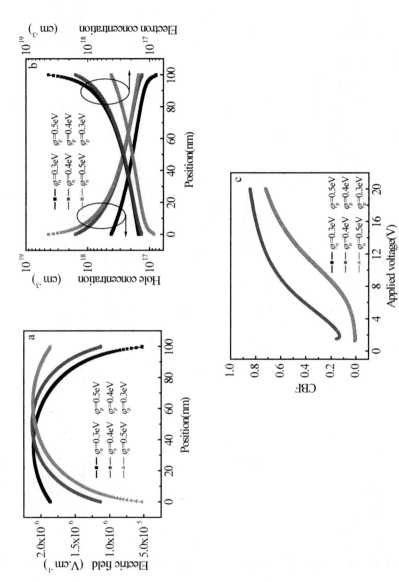

FIG. 2 Calculated distributions of (a) the electric field intensity; (b) hole and electron concentration in bulk of device; and (c) the relationship between applied voltage and CBF value. The parameters used are: $\mu_n = 5 \times 10^{-5} \text{cm}^2 \cdot V^{-1} \cdot s^{-1}$, $\mu_p = 5 \times 10^{-5} \text{cm}^2 \cdot V^{-1} \cdot s^{-1}$.

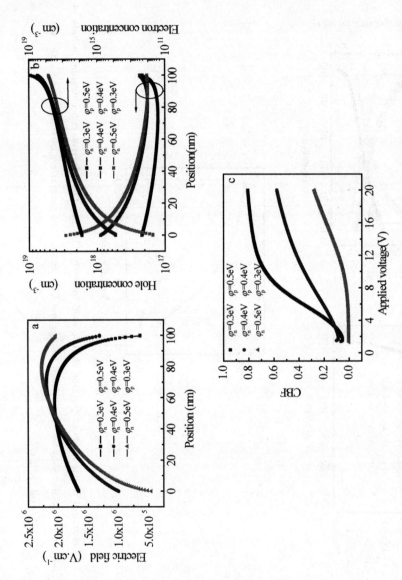

FIG. 3 Calculated distributions of (a) the electric field intensity;(b) hole and electron concentration in bulk of device;and (c) the relationship between applied voltage and CBF value. The parameters used are:$\mu_n = 1 \times 10^{-6}$ cm^2·V^{-1}·s^{-1}, and $\mu_p = 1 \times 10^{-5}$ cm^2·V^{-1}·s^{-1}.

3.3 $\mu_{0n} < \mu_{0p}$

For a single layer OLED with $\mu_{0n} < \mu_{0p}$, Fig. 3 shows the electric field intensity (Fig. 3-a), the hole and the electron concentration (Fig. 3-b) distribution in the bulk of organic layer, and the relationship between applied voltage and CBF value (3-c) under the condition that $\varphi_n > \varphi_p$, $\varphi_n = \varphi_p$, and $\varphi_n < \varphi_p$. Fig. 3-c shows that devices can achieve the largest CBF value when $\varphi_n = \varphi_p$ in the lower applied voltage region, and can achieve the largest CBF value in the case of $\varphi_n > \varphi_p$ in the higher applied voltage region. The CBF value increases as the applied voltage increases, similar to the case when $\mu_{0n} > \mu_{0p}$.

4 CONCLUSION

The influence of applied voltage on the CBF value of single layer OLEDs was studied based on the numerical model of a bipolar single layer OLED with organic layer trap free and without doping. The results showed that in cases when $\mu_{0n} >$, =, and $< \mu_{0p}$ in the lower applied voltage region, the largest CBF can be still be achieved when $\varphi_n = \varphi_n$. In the higher voltage region, the largest CBF can be achieved in the case of $\varphi_n > \varphi_p$ under the condition of $\mu_{0n} > \mu_{0p}$, with the result being the reverse in the case of $\mu_{0n} < \mu_{0p}$. When $\mu_{0n} = \mu_{0p}$, the largest CBF can be achieved in the case of $\varphi_n = \varphi_p$ in the entire region of the applied voltage. In addition, the CBF value of the device increases with increasing applied voltage.

ACKNOWLEDGMENTS

The work leading to these results has received funding from the National Natural Science Foundation of China (Grants Nos. 11265013, 11264033), The Science Research Foundation of Tianshui Normal University (Grant No. TSA1108).

Peferences

[1] S. Reineke, F. Lindner, G. Schwartz, N. Seidler, K. Walzer, B. Lussem, and K. Leo, Nature 459, 234 (2009).

[2] P. Freitag, S. Reineke, S. Olthof, M. Furno, B. Lüssem, and K. Leo, Org. Electron. 11, 1676 (2010).

[3] M. C. Gather, A. Köhnen, and K. Meerholz, Adv. Mater. 23, 233 (2011).

[4] G. L. Mao, Z. X. Wu, Q. He, B. Jiao, G. j. Xu, X. Hou, Z. J. Chen, and Q. H. Gong, Appl. Surf. Sci. 257, 7394 (2011).

[5] S. A. Van Slyke, C. H. Chen, and C. W. Tang, Appl. Phys. Lett. 69, 2160 (1996).

[6] J. X. Luo, L. X. Xiao, and Q H Gang, Appl. Phys. Lett. 93, 133301 (2008).

[7] L. S. Hung, C. W. Tang, and M. G. Mason, Appl. Phys. Lett. 70, 152(1997).

[8] F. Li, H. Tang, J. Anderegg, and J. Shinar, Appl. Phys. Lett. 70, 1233(1997).

[9] X. W. Zhang, J. W. Xu, H. R. Xu, H. P. Lin, J. Li, X. Y. Jiang, and Z. L. Zhang, Optics & Laser Technology 45, 181(2013).

[10] S. M. Tadayyon, H. M. Grandin, K. Griffiths, P. R. Norton, H. Aziz, and Z. D. Popovic, Org. Electron. 5, 157(2004).

[11] H. You, Y. F. Dai, Z. Q. Zhang, and D. G. Ma, J. Appl. Phys. 101, 026105(2007).

[12] G. G. Malliaras, J. R. Salem, P. J. Brock, and C. Scott, Phys. Rev. B, 1998, 58(20):R13411.

[13] I. M. Chan and F. C. Hong, Thin Solid Films 450, 304(2004).

[14] J. P. J. Markham, S. - C. Lo, S. W. Magennis, P. L. Burn, and I. D. W. Samuel, Appl. Phys. Lett. 80, 2645(2002).

[15] P. W. M. Blom, M. J. M. de Jong, and S. Breedijk, Appl. Phys. Lett. 71, 930(1997).

[16] P. S. Davids, I. H. Campbell, and D. L. Smith, J. Appl. Phys. 82, 6319(1997).

[17] A. B. Walker, A. Kambili, and S. J. Martin, J Phys – Condens Mat. 14, 9825(2002).

[18] P. S. Davids, I. H. Campbell, and D. L. Smith, J. Appl. Phys. 82(12), 6319, (1997).

[19] R. H. Fowler and L. Nordheim, Proc. R. Soc. London, Ser. A 119, 173(1928).

[20] H. K. Gummel, IEEE Trans. Electron Dev. ED – 11 455(1964).

[21] D. L. Scharfetter and H. K. Gummel, IEEE Trans. Electron Dev. ED 16, 64(1969).

注:本文曾发表在2014年4月《Journal of Applied Physics》第16期上

Numerical model of tandem organic light-emitting diodes based on transition metal oxide interconnector layer

Lu Feiping　Peng Yingquan　Xing Yongzhong*

摘　要：本论文采用两步法去描述过渡金属氧化物薄膜连接层的电荷产生与分离机制,建立了基于过渡金属氧化物薄膜为连接层的叠层有机电致发光器件的数值模型。该型不仅适用于p型过渡金属氧化物,也适用于n型过渡金属氧化物。基于该模型,首先研究了注入势垒对连接层电荷产生能力的影响,其次研究了器件内电场强度、载流子浓度以及电势的分布。研究结果表明,当保持器件一侧注入势垒不变,逐渐升高另一侧注入势垒时,载流子的注入能力降低,连接层的电荷产生能力也逐渐降低,注入界面处电场强度逐渐增强,电场分布逐渐趋于线性,两个发光单元中的电势降落趋于一致,且计算结果与实验结果吻合的很好。本文获得的结果可为深入了解基于过渡金属氧化物薄膜为连接层的叠层有机电致发光器件的工作机理提供参考。

By utilizing a two – step process to express the charge generation and separation mechanism of transition metal oxide(TMOs) interconnector layer, a numerical model was proposed for the tandem organic light emitting diodes(OLEDs) with TMOs thin film as interconnector layer. This model was valid not only for n – typed TMOs interconnector layer, but also for n – typed TMOs interconnector layer. Based on this model, the influences of different carrier injection barrier at the interface of electrode/organic layer on the charge generation ability of interconnector layers were studied. Besides, the distribution characteristics of carrier concentration, electric field intensity and potential in de-

---

* 作者简介:路飞平(1980—　),男,甘肃岷县人,天水师范学院副教授、博士,主要从事有机光电子材料及器件物理研究。

vice under different carrier injection barrier were studied. The results show that when keeping one carrier injection barrier as a constant and increasing another carrier injection barrier, carriers injected into device will be gradually decreased, and the carrier generation ability of interconnector layer were gradually reduced, the electric field intensity at the interface of organic/electrode will be gradually enhanced, and the electric field distribution became nearly linearly, the voltage drop in two light units were gradually became the same. Meanwhile, the carrier injection ability will be decreased as another carrier injection barrier increasing. The obtained results can give us a deep understanding of work mechanism of TMOs – based tandem OLEDs. The simulation results can agree with the experimental data.

1. Introduction

Due to the great development achieved in organic light emitting diodes(OLEDs), OLEDs have been widely used in flat – panel displays and solid state lighting [1-6]. In general, a long operating lifetime must be ensured before mass – production of OLEDs for practical use. Tandem OLEDs having two or multiple electroluminescence units vertically stacked in series through interconnector layers can meet this requirement, because they have the advantages of enhanced current efficiency and luminance at a relative low current density, the lifetime of devices can be prolonged as compared to conventional single – unit devices[7-14]. In a tandem OLED, the interconnector layer playing an important role that serves as the charge generation layer, are the critical factor in performance of tandem OLEDs. Upon the application of an applied voltage, the interconnector layers will generate electrons and holes. The generated electrons and holes will be injected into the neighboring electron – and hole – transporting layers(ETL and HTL) of the individual light – emitting units, and recombine with holes injected from the anode side or electrons injected from the cathode side for light emission.

Transition metal oxides(TMOs), such as molybdenum tri – oxide(MoO_3)[11,15,16], tungsten trioxide(WO_3)[17,18], vanadium oxide V_2O_5[10,19], because they have high light transition, high work function and can be thermally deposited, were widely used as the interconnector layer to fabricate high efficiency tandem OLEDs. Up to now, several models have been proposed to interpret the charge generation and separation process in the TMO – based tandem OLEDs. Such as thermally assisted charge generation mode[19,20], in this model, the TMOs were considered as the p – typed semiconductors, the electrons transfer from valence band(VB) of TMOs to the impurity level via thermionically excited, and then tunneling through a triangular energy barrier by electric field assistance. The triangular energy barrier is the difference between the impurity level and the lowest unoccupied molecular orbitals(LUMO) of the n – doped organic layer adjacent to TMO interconnector layer. Besides, according to the electronic structures determined by photoemission spectroscopy (UPS) analysis, TMOs actually show n – type semiconducting property with high work function and a deep – lying conduction band, charge generation and separation occur at the interface between TMOs and the adjacent HTL via electron transfer from the HTL's highest occupied molecular orbital(HOMO) into the conduction band(CB) of TMOs[21-25]. Recently, Yang et al[26], by systematically investigating electrical and spectral emission properties, interface energetic, and capacitance characteristics, show that vacuum – deposited MoO_3 layer acts as a charge generation layer due to spontaneous electron transfer from various defect states to the conduction band via thermal diffusion. The external electric – field induces the charge separation via injection of electrons and holes into the neighboring n – doped organic layer and hole – transporting layers, respectively. However, there is little study about carrier concentration, electric field and potential distribution characteristics in tandem OLEDs, and the relationship between the charge generation ability and injection barrier height of electrode/organic layer is not studied. In this paper, we proposed a numerical model for TMOs – based tandem OLEDs. In this model, charge generation and separation process was expressed with a two – step process, and the model was valid not only for the n – typed TMOs but also for the p – typed TMOs. For the TMOs layer are n – typed semiconductors, the charge generation process is that the electrons transfer from various gap state level(ϕ_t) to the CB of TMOs via thermionically excited, leaving the holes at the gap state. The generated electrons and holes will tunnel through a triangular energy barrier by electric field assistance injecting into light units. For electrons, the triangular energy barrier (ϕ_{B1}) is the difference between the CB of TMOs and the LUMO of the organic layer adjacent at left side of TMO layer. For holes, the triangular energy barrier (ϕ_{B2}) is the difference between gap state level of TMOs and the HOMO of the organic layer at the right side of TMO layer, seeing from figure 1. The intensity of generated current density is determined by the larger one of ϕ_{B1} and ϕ_{B2}. For the TMOs is p – typed semiconductors, the charge generation and separation process is similar with the model proposed by Qi[20], in this case, the triangular energy barrier(ϕ_{B1}) for electrons is the difference of the impurity level of TMOs and the LUMO of the organic layer adjacent at left side of TMO layer, for holes, the triangular energy barrier (ϕ_{B2}) is the difference between the VB of TMOs and the HOMO of the organic layer at the right side of TMO layer. Based on this model, the carrier concentration, electric field and potential distribution characteristics under different injection barrier were investigated. Besides, the influences of carrier injection barrier at the interface of electrode/organic layer on the charge generation ability were discussed. Also, the simulation results can agree with the experimental data. The obtained results can give us a deep understanding of work mechanism for TMOS – based tandem OLEDs.

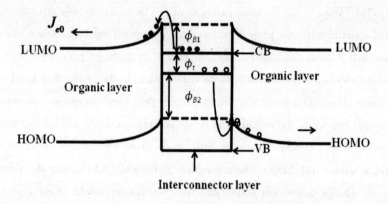

Figure 1. Diagram of charge generation process in TMO – based interconnector layer

2. Numerical model

The TMO – based tandem OLEDs used here is that the tandem OLEDs with two identical light units connected by a TMOs thin film, and the light unit is a single organic layer. We define the position of hole injection at anode is $x = 0$, and electron injection at cathode is $x = d$, d is the total thickness of organic layers, seeing from figure 2. In general, TMOs film can be prepared via thermal evaporation, however, high prepared temperature leads to oxygen releasing from TMOs film, some transition metal atoms at the film surface are reduced, with the missing oxygen atoms. The partly filled d orbitals of reduced transition metal atoms are result in gap states[27], which can be found in UPS measurements for MoO_3[28], WO_3 film[29]. Furthermore, the reduced TMOs film will increase the conductivity[27]. Because the ϕ_t of TMO layer is very small, about 0.5 eV[20,27,28], and the gap state density is high, such as can be 1×10^{18} cm^{-3}[20], it will result in TMO thin layer with higher conductivity comparing with the organic materials. For this reason, in this model, in order to simplify the calculation process, we ignore the voltage drop on the TMOs thin layer, and considered it to be an ideal surface. Upon the application of an applied voltage, TMOs – based tandem OLEDs will function, the generated current density of electron(J_{e0}) and hole(J_{h0}) are shown in Figure 2. Similar to the Qi's model, the generation current density can be expressed by a two – step process[20]:

$$J_{e0} = J_{h0} = qv_e N_t f P(V), \quad (1)$$

here, q is the elementary charge, v_e is the electron thermal velocity, $v_e = (\frac{kT}{m_e^*})^{\frac{1}{2}}$, N_t is the gap states concentration in TMOs thin film, f is the Fermi – Dirac function, $f = \frac{1}{1 + \exp(q\varphi_t/kT)}$, k is Boltzmann's constant, T is the working temperature, and $P(V)$ is the tunneling probability over an interface barrier of height ϕ_B, can be expressed as:

$$P(V) = \exp(-\frac{\alpha}{E(V)}\varphi_B^{\frac{3}{2}}), \quad (2)$$

here ϕ_B is the larger one of ϕ_{B1} and ϕ_{B2}, seeing from figure 2,

$$\varphi_B = \begin{cases} \varphi_{B1}, & \varphi_{B1} > \varphi_{B2} \\ \varphi_{B2}, & \varphi_{B1} < \varphi_{B2} \end{cases} \quad (3)$$

For a triangular energy barrier, $\alpha = \frac{8\pi\sqrt{2m_e^* q}}{3h}$, $E(V)$ is the electric field in the TMOs thin film interconnector at voltage V, m_e^* is the electron effective mass in the organic semiconductor, and h is the Planck's constant.

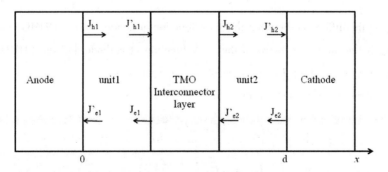

Figure 2. The schematic of the injection and leakage current density for holes(J_h, J'_h) and electrons(J_e, J'_e) of both light emitting units.

For a tandem OLED, the components of device current density process are shown in the figure 2. J_{h1} and J_{e2} is the hole current density and electron current density injected from anode and cathode. The electrode/organic interfaces always is Schottky contacts, there are three components to J_{h1} and J_{e2}: tunneling current, thermionic emission current, and a back flowing interface recombination current that is the time reversed process of thermionic emission[30,31], so they can be expressed as:

$$J_{h1} = J_{1tu} + J_{1th} - J_{1ir}, \qquad (4)$$

$$J_{e2} = J_{2tu} + J_{2th} - J_{2ir}, \qquad (5)$$

where $J_{1tu}(J_{2tu})$, $J_{1ir}(J_{2ir})$ and $J_{1th}(J_{2th})$ is the tunneling current, thermionic emission current and back flowing interface recombination current of hole (electron) injection at the interface of anode (cathode)/organic layer. The tunneling current density can be expressed as[32]:

$$J_{1tu} = \frac{C}{\varphi_{h,eff}} F(0) \exp\left(-\frac{B\sqrt{\varphi_{h,eff}^3}}{F(0)}\right), \qquad (6)$$

$$J_{2tu} = \frac{C}{\varphi_{n,eff}} F(L) \exp\left(-\frac{B\sqrt{\varphi_{n,eff}^3}}{F(L)}\right), \qquad (7)$$

where
$$C = \frac{2.2q^3}{8\pi h}, \quad B = \frac{8\pi\sqrt{2m^*}}{2.96h}$$

here $F(0)$ and $F(L)$ is the electric field intensity at the interface of anode/organic and cathode/organic, respectively, h is Planck constant. $\phi_{h,eff}$ and $\phi_{n,eff}$ is the effective interfacial energy barrier for anode/organic and cathode/organic, can be expressed as:

$$\varphi_{h,eff} = \varphi_h - q\sqrt{\frac{q|F(0)|}{4\pi\varepsilon}} \text{ and } \varphi_{n,eff} = \varphi_e - q\sqrt{\frac{q|F(L)|}{4\pi\varepsilon}}, \qquad (8)$$

here ϕ_h, is the difference between the work function of anode and the HOMO of organic layer, ϕ_e, is the difference between the work function of cathode and the LUMO of organic layer, ε is the dielectric constant of organic materials.

The thermionic emission current can be expressed as:

$$J_{1th} = AT^2 \exp\left(-\frac{\varphi_{h,eff}}{kT}\right), \qquad (9)$$

$$J_{2th} = AT^2 \exp\left(-\frac{\varphi_{e,eff}}{kT}\right), \qquad (10)$$

here, A is Richardson's constant. The back flowing interface recombination current can be expressed as[33]:

$$J_{1ir} = v_p p(0) , \quad J_{2ir} = v_e n(L) , \tag{11}$$

here, $p(0)$ and $n(0)$ is the carrier density for holes and electrons at the interface of electrode/organic, v_p and v_e is the kinetic coefficient determined by detailed balance between thermionic emission and interface recombination which are time reversed processes of each other:

$$v_p = \frac{AT^2}{p_0} , \quad v_e = \frac{AT^2}{n_0} , \tag{12}$$

here p_0 and n_0 is the density of ways that can be occupied by a hole or electron.

At the interface of organic/TMOs thin film, there also exist interface recombination current, expressed as:

$$J_e^- = v_e n^{(L)} , \quad J_h^+ = v_e p^{(R)} \tag{13}$$

here $n^{(L)}$ is the electron concentration at the left side of interface of organic/TMO, $p^{(R)}$ is the hole concentration at the right side of organic/TMO. So the injection current density of J_{e1} and J_{h2} (show in figure 2) can be expressed as:

$$J_{e1} = J_{e0} + J'_{e2} - J_e^- , \tag{14}$$

$$J_{h2} = J_{h0} + J'_{h1} - J_p^+ , \tag{15}$$

here, J'_{h1} and J'_{e2} is the hole leakage current from the light unit 1 and electron leakage current from the light unit 2, respectively.

The time-independent current continuity equations for electron and hole in device are:

$$\frac{dJ_p}{dx} = q(G - R) \tag{16}$$

$$\frac{dJ_n}{dx} = -q(G - R) \tag{17}$$

G and R is the carrier generation and recombination rate, because of the high energy band for organic materials, generation rate is very small and can be ignored. Assuming that the recombination to be Langevin type, then we have[34]:

$$R = \frac{4\pi q}{\varepsilon}(\mu_n + \mu_p)(np - n_i^2) \tag{18}$$

here ε_0 is the permittivity of vacuum, ε_r is the relative dielectric constant of the organic materials, n and p are the electron and hole density respectively, μ_n and μ_p are the electron and hole mobility respectively, always are filed-dependent:

$$\mu_n(E) = \mu_{0n} \exp\left(\sqrt{\frac{F}{F_{0n}}}\right) \tag{19}$$

$$\mu_p(E) = \mu_{0p}\exp(\sqrt{\frac{F}{F_{0p}}}) \qquad (20)$$

here μ_{0n} and μ_{0p} are the zero field mobility, F_{0n} and F_{0p} are the characteristic electric fields for electrons and holes, respectively, F is the electric field, n_i is the intrinsic carrier density in the organic semiconductor defined by:

$$n_i^2 = N_{HOMO}N_{LUMO}\exp(-\frac{E_g}{k_B T}) \qquad (21)$$

here N_{HOMO} and N_{LUMO} are the total densities of states of HOMOs and LUMOs. E_g is the energy gap between LUMO and HOMO, k_B is Boltzmann constant. In the device bulk, hole and electron current density can be expressed as:

$$J_p = qp\mu_p F - qD_p \frac{\partial p}{\partial x} \qquad (22)$$

$$J_n = qn\mu_n F + qD_n \frac{\partial n}{\partial x} \qquad (23)$$

here D_p and D_n are the diffusion coefficients of holes and electrons, respectively. n and p is electron and hole concentration, respectively. Assuming the Einstein relation is still valid, we have:

$$\frac{D_n(F)}{\mu_n(F)} = \frac{D_p(F)}{\mu_p(F)} = \frac{k_B T}{q} \qquad (24)$$

The device current density J is the sum of J_n and J_p,

$$J = J_n + J_p \qquad (25)$$

Under the steady state, from the principle of current continuality, we can get the following relations:

$$J(x) = J_{h1} + J'_{e1} = J_{e1} + J'_{h1} = J_{h2} + J'_{e2} = J_{e2} + J'_{h2} \qquad (26)$$

Also we can have:

$$\frac{dJ(x)}{dx} = 0 \qquad (27)$$

The boundary condition for Poisson's equation is:

$$V - V_{built-in} = \int_0^L F(x)dx \qquad (28)$$

here L is the total thickness of two light units.

The electric field F and electrostatic potential V are coupled with n and p through Poisson equation:

$$\frac{\partial^2 V}{\partial x^2} = -\frac{\partial F}{\partial x} = \frac{q}{4\pi\varepsilon}(p - n) \qquad (29)$$

The above set of equations is solved numerically by the Scharfetter – Gummel method [35,36].

3. Results and discussion

For an ideal single layer organic light – emitting diode, if the mobility of holes and electrons are equal, the recombination and emission zone can be located in the centre of light units, OLEDs can have a high light efficiency[34]. So in this paper, we mainly study the property of the tandem OLEDs under the condition that the emitting layer with same mobility for holes and electrons. The parameters used in simulation are showed in table 1.

Table 1 The parameters used in simulation

E_g (eV)	3.0	N_{LUMO} (cm^{-3})	1×10^{21}
μ_{n0} (cm^2·V^{-1}·s^{-1})	1×10^{-5}	N_{HOMO} (cm^{-3})	1×10^{21}
μ_{p0} (cm^2·V^{-1}·s^{-1})	1×10^{-5}	ϕ_t (eV)	0.5
F_{n0} (V·cm^{-1})	5×10^5	ϕ_B (eV)	1.0
F_{p0} (V·cm^{-1})	5×10^5	n_0 (cm^{-3})	1×10^{21}
T(k) 300	p_0 (cm^{-3})	1×10^{21}	
N_t (cm^{-3})	1×10^{15}	V	12

3.1 Keep the hole injection barrier ϕ_h a constant, varying electron injection barrier ϕ_e

Figure 3 and figure 4 show the electron and hole concentration distribution in device under different electron injection barrier ϕ_e. We can see from the figure 3, when keeping the hole injection barrier ϕ_h as a constant(=0.5 eV), and increasing the electron injection barrier ϕ_e, the number of electrons injected into organic layer from cathode are gradually reduced. Meanwhile, we can see from figure 3 the electron concentration on the left side TMO interconnector layer is also reduced, which is mainly contributed by the interconnector layer. It tell us that the carrier generation ability is reduced as the increasing of ϕ_e. Although the hole injection barrier ϕ_h is kept as a constant, the hole injection ability from anode is also reduced. We can see it from the hole concentration distribution in the vicinity of the anode in figure 4. As one know, upon the application of an applied voltage, electrons and holes will be injected into organic layer. At the same time, TMOs interconnector layers will generate carriers and inject them into two light units respectively via tunneling – assisted thermionic emission. When the injection barrier ϕ_e is increased, the number of electrons injected from cathode into organic layer will be reduced, it will lead to the carriers recombination in light unit 2 reduced, and the generation ability of interconnector layer will be restrained. However, if the injection barrier ϕ_e is decreased, more electrons will be injected from cathode into light unit 2, and they will quickly combine with the holes generated by the interconnector layer, the recombination ability is improved comparing with the case of large ϕ_e. In this case, the hole generation ability of the interconnector layer will be improved. On the other hand, the holes will be injected from anode into light unit under an applied electric field, because of more electrons generated by the interconnector layer, the carrier combination in light unit 1 will be improved, this will result in more holes being injected from anode into light unit 1, seeing from figure 4, the hole concentration at the interface of anode is increased.

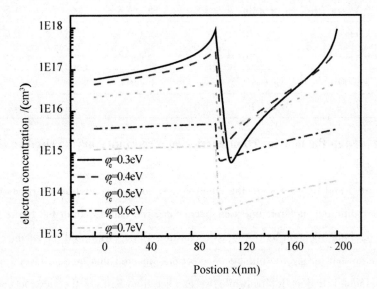

Figure 3 The electrons distribution in tandem OLED with different electron injection barrier

Figure5 is the electric field distribution in organic layer, its distribution characteristics is determined by the carrier distribution in the device according to the equation (29). As the increasing of ϕ_e, the electric field distribution is approach to lineally, and the electric field intensity in the vicinity of cathode interface is gradually enhanced. Because of low carrier mobility in organic materials, a lot of injected electrons will accu-

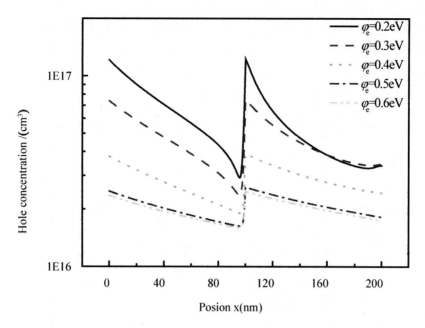

Figure 4　The holes distribution in tandem OLED with different electron injection barrier

mulate in the vicinity of cathode interface, the accumulated electrons will produce an inner electric field with opposite direction comparing with the applied electric field. The produced inner electric field will reduce the electric field intensity in the device bulk. The more electrons were injected, the more electrons will be accumulated in the vicinity of cathode interface, and so the intensity of electric field in there will be weaker, seeing from figure 5. When the ϕ_e is little, such as $\phi_e = 0.3$ eV, in this case, more electrons were injected into devices from cathode. In some regions, the majority carrier is hole, and in other regions is electron. In the region of hole concentration($p(x)$) larger than electron concentration($n(x)$), such as in region $x_A < x < x_B$ and $x_C < x < x_D$, the slope of electric field distribution curve is larger than 0, electric field intensity is increasing gradually. In the region of $p(x) < n(x)$, such as in region $x_B < x < x_C$ and $x_D < x < x_E$, the slope of electric field distribution curve is less than 0, electric field intensity is decreasing gradually. At the position of $p(x) = n(x)$, such as at the position $x = x_B, x = x_C$ and $x = x_D$, the slope of electric field distribution curve is 0(seeing from figure 6). Also we can see from figure 6, at the region where carrier concentration vary sharply, the absolute value of the slope of electric field distribution curve also vary sharply, such as in the vicinity of position x_C, x_E. When the ϕ_e is large, such as $\phi_e = 0.7$ eV, in this case, electrons injected into device from cathode is greatly reduced comparing with $\phi_e = 0.3$ eV. In the whole region of device bulk, because of the $p(x) > n(x)$, the slop of electric field distribution curve is larger than 0, the electric field intensity is gradually increasing from anode to cathode, seeing from figure 7.

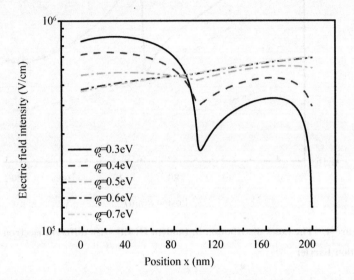

Figure 5 The electric field distribution in tandem OLED with different electron injection barrier

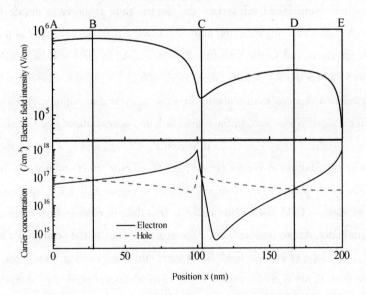

Figure 6 The carrier concentration and electric field distribution in tandem OLED with different hole injection barrier, $\phi_e = 0.3$ eV, $\phi_h = 0.5$ eV.

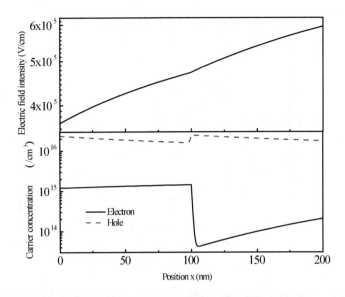

Figure 7 The carrier concentration and electric field distribution in tandem OLED with different hole injection barrier, $\phi_e = 0.7$ eV, $\phi_h = 0.5$ eV.

Figure8 is the potential distribution in organic layer. as the increasing of ϕ_e, the potential distribution is become lineally, this characteristic is also determined by the carriers distribution in the devices according to equation(29). One can see from the figure, when the ϕ_e is little, the voltage drop in the light unit1 is larger than in the light unit2. As the increasing of ϕ_e, the voltage drop in both of light unit will become nearly same.

3.2 Keep the electron injection barrier ϕ_e a constant, varying hole injection barrier ϕ_h

Figure9 and figure 10 is the electron and hole concentration distribution in the device, keeping the electron injection barrier ϕ_e is constant($=0.5$ eV), and gradually increasing ϕ_h. In this case, have the same reason(expressed in section 3.1), as the hole injection barrier ϕ_h increasing, the hole injection ability is reduced, and the carrier generation ability of interconnector layer is reduced, it lead to the electron injection ability from cathode is reduced.

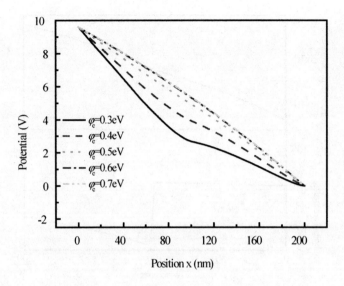

Figure 8 the potential distribution in tandem OLED with different electron injection barrier

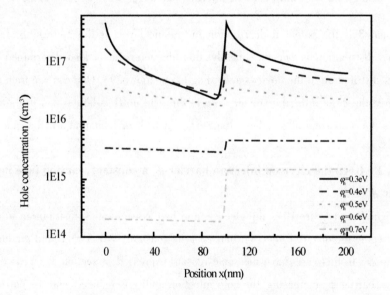

Figure 9 the holes distribution in tandem OLED with different hole injection barrier

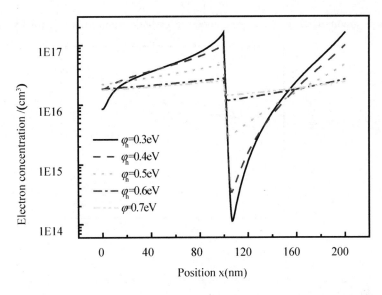

Figure 10 The electrons distribution in tandem OLED with different hole injection barrier

Figure11 is the electric field distribution in the device, while keeping electron injection barrier ϕ_e as a constant ($=0.5$ eV) and increasing the hole injection barrier ϕ_h. As the increasing of ϕ_h, the electric field distribution is approach to lineally, and the intensity of electric field in the vicinity of anode is gradually enhanced. When the ϕ_e is little, such as $\phi_h = 0.3$ eV, in this case, more holes were injected into devices from anode. In some region, the majority carrier is hole, and in other region is electron. In the regions of $p(x) > n(x)$, such as in region $x_A < x < x_B$ and $x_C < x < x_D$, the slope of electric field distribution curve is larger than 0, electric field intensity is gradually increasing in those regions. In the regions of $p(x) < n(x)$, such as in region $x_B < x < x_C$ and $x_D < x < x_E$, the slope of electric field distribution curve is less than 0, electric field intensity is gradually decreasing in those regions. At the position of $p(x) = n(x)$, such as at the position $x = x_B, x = x_C$ and $x = x_D$, the slope of electric field distribution curve is 0 (seeing from figure 12). Also we can see from figure 12, at the region of carrier concentration varying sharply, the absolute value of the slope of electric field distribution curve is also varying sharply, such as in the vicinity of position of x_A and x_C. When the ϕ_h is large, such as $\phi_h = 0.7$ eV, in this case, holes injected into device from anode is reduced comparing with $\phi_h = 0.3$ eV. In the whole region of device bulk, because of the $p(x) < n(x)$, the slop of electric field distribution curve is less than 0, the electric field intensity is gradually decreasing from anode to cathode seeing from figure 13.

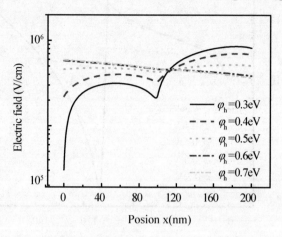

Figure. 11　The electric field distribution in tandem OLED with different hole injection barrier

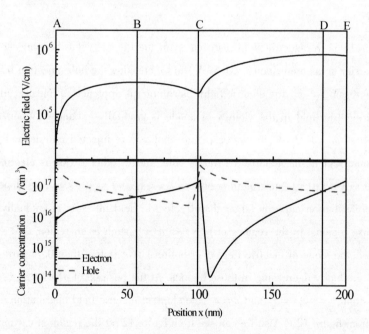

Figure 12　The carrier concentration and electric field distribution in tandem OLED with different hole injection barrier, $\phi_e = 0.5$ eV, $\phi_h = 0.3$ eV.

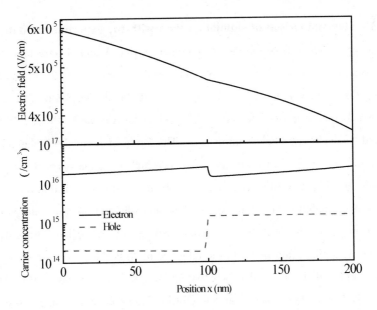

Figure 13 The carrier concentration and electric field distribution in tandem OLED with different hole injection barrier, $\phi_e = 0.5$ eV, $\phi_h = 0.7$ eV.

Figure 14 is the potential distribution in the device, keeping electron injection barrier ϕ_e as a constant ($=0.5$ eV), increasing the hole injection barrier ϕ_h. The distribution characteristic is also determined by the carrier distribution in devices. In this case, when the ϕ_h is little, the voltage drop in the light unit 1 is lessen than in the light unit 2. As the ϕ_h increasing, the voltage drop in both light units is became nearly same.

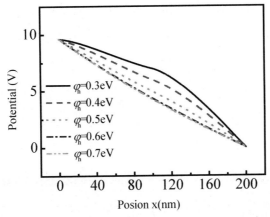

Figure 14 The potential distribution in tandem OLED with different hole injection barrier

97

3.3 The comparison of simulation data with the experimental data

In order to demonstrate the model is proper, device A and B were fabricated. The devices structure is glass/ITO/Alq$_3$(80 nm)/MoO$_3$(x nm)/ NPB(40 nm)/LiF(2 nm)/Al(100 nm), with $x = 0$ for device A, $x = 10$ for device for B. The fabrication process can be found in the reference of [16], here the Alq$_3$ and NPB thin film can be considered as single layer light emitting unit. The figure 15 is the $J - V$ characteristic of device A and device B, As we all know, because high injection barrier (holes injected from anode into Alq$_3$ thin film layer and electrons injected into NPB thin film layer) and very low carrier mobility (hole transport in Alq$_3$ thin film and electron transport in NPB thin film), devices A has very little current density under an applied voltage. However, when inserting a MoO$_3$ thin film into the interface of NPB/Alq$_3$, the current density can be greatly improved. Those results show that MoO$_3$ thin film is an essential to the charge generation and separation process tandem OLEDs. Figure 16 is the comparison of simulation result with the experiment data, when use the reasonable parameters, we can see the experiment result can agree well with the simulation results. It indicates that this numerical model can be used to express the work mechanism of the tandem OLEDs based on TMOs interconnector layer. The parameters used in simulation are showed as following. For Alq$_3$, the parameters used in simulation is: $E_g = 2.6$ eV, $\mu_{n0} = 3 \times 10^{-6}$ cm^2·V^{-1}·s^{-1}, $\mu_{p0} = 8 \times 10^{-8}$ cm^2·V^{-1}·s^{-1}, $F_{n0} = 2 \times 10^5$ V·cm^{-1}, $F_{p0} = 2 \times 10^5$ V·cm^{-1}, $N_{LUMO} = 1 \times 10^{21}$ cm^{-3}, $n_0 = 1 \times 10^{21}$ cm^{-3}, $p_0 = 1 \times 10^{21}$ cm^{-3}. For NPB, $E_g = 3$ eV, $\mu_{n0} = 6 \times 10^{-8}$ cm^2·V^{-1}·s^{-1}, $\mu_{p0} = 6 \times 10^{-4}$ cm^2·V^{-1}·s^{-1}, $F_{n0} = 2 \times 10^5$ V·cm^{-1}, $F_{p0} = 2 \times 10^5$ V·cm^{-1}, $N_{LUMO} = 1 \times 10^{21}$ cm^{-3}, $n_0 = 1 \times 10^{21}$ cm^{-3}, $p_0 = 1 \times 10^{21}$ cm^{-3}. And the $\phi_t = 0.5$ eV, $\phi_B = 2.5$ eV, $N_t = 1 \times 10^{18}$ cm^{-3}.

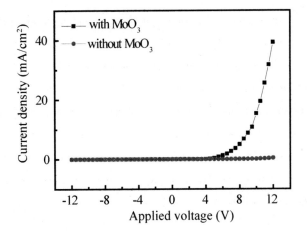

Figure 15 The J – V characteristic of device A and B

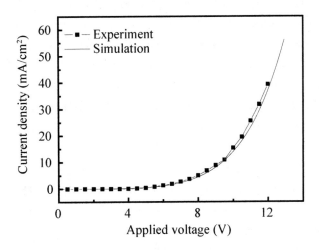

Figure 16 The comparison of simulation results with experimental results

4. Conclusion

In summary, by utilizing a two – step process charge generation mechanism for transition metal oxide interconnector layer, which consisting of tunneling – assisted thermionic emission over an injection barrier and agap state below the transition metal oxide conduction band edge, a numerical model was proposed for the TMO – based tandem OLEDs. The distribution characteristic of carrier density, electric field and potential in tandem OLEDs under different carrier injection barrier were investigated. When keeping the one carrier injection barrier as a constant and increase another carrier injection barrier, carriers injected into device will be gradually decreased, and the carrier generation ability of interconnector layer were gradually reduced, the electric field intensity at the interface of organic/electrode will be gradually enhanced, and the electric field distribution became nearly linearly, the voltage drop in two light units were gradually became the same. Meanwhile, the carrier injection ability of another electrode was decreasing as the carrier injection barrier increasing. Also, the experimental data can agree with the simulation results. Those results can give us a deep understanding of work mechanism of tandem OLEDs.

References

[1] Reineke S, Lindner F, Schwartz G, et al. White organic light – emitting diodes with fluorescent tube efficiency. Nature, 2009, 459(7244):234

[2] Zhang Y W, Chen W B, Liu H H. A new AC driving circuit for a top emission AMOLED. Chinese Journal of Semiconductors, 2013, 34(5):055005 – 5

[3] Wang H, Wang ZG, Feng J, et al. A pixel circuit with reduced switching leakage for an organic light – emitting diode. Chinese Journal of Semiconductors, , 2012, 33(12):125006 – 5

[4] Zhao B H, Huang R, Bu J H, et al. A new OLED SPICE model for pixel circuit simulation in OLED – on – silicon microdisplay design. Chinese Journal of Semiconductors, 2012, 33(7): 075007 – 6

[5] Zhao B H, Huang R, Ma F, et al. The realization of an SVGA OLED – on – silicon microdisplay driving circuit. Chinese Journal of Semiconductors, 2012, 33(3):035006 – 6

[6] Huang R, Wang X H, Wang W B, et al. Design of a 16 gray scales 320 × 240 pixels OLED – on – silicon driving circuit. Chinese Journal of Semiconductors, 2009, 30(1):015010 – 4

[7] Chiba T, Pu Y J, Miyazaki R, et al. Ultra – high efficiency by multiple emission from stacked organic light – emitting devices. Organic Electronics, 2011, 12(4):710 – 715.

[8] Liao L S, Slusarek W K, Hatwar T K, et al. Tandem Organic Light – Emitting Diode using Hexaazatriphenylene Hexacarbonitrile in the Intermediate Connector. Advanced Materials, 2008, 20 (2):324

[9] Chen C W, Lu Y J, Wu C C, et al. Effective connecting architecture for tandem organic light – emitting devices. Applied Physics Letters, 2005, 87(24):241121

[10] Tsutsui T, Terai M. Electric field – assisted bipolar charge spouting in organic thin – film diodes. Applied physics letters, 2004, 84(3):440

[11] Kanno H, Holmes R J, Sun Y, et al. White Stacked Electrophosphorescent Organic Light – Emitting Devices Employing MoO_3 as a Charge – Generation Layer. Advanced Materials, 2006,

18(3):339.

[12] Chen P, Xue Q, Xie W, et al. Influence of interlayer on the performance of stacked white organic light – emitting devices. Applied Physics Letters, 2009, 95(12):123307.

[13] Guo F, Ma D. White organic light – emitting diodes based on tandem structures. Applied Physics Letters, 2005, 87(17):173510.

[14] Liao L S, Klubek K P, Tang C W. High – efficiency tandem organic light – emitting diodes. Applied physics letters, 2004, 84(2):167.

[15] Meyer J, Kröger M, Hamwi S, et al. Charge generation layers comprising transition metal – oxide/organic interfaces: Electronic structure and charge generation mechanism. Applied Physics Letters, 2010, 96:193302.

[16] Lu F P, Wang Q, Zhou X. Tandem organic light – emitting diode with a molybdenum tri – oxide thin film interconnector layer. Chinese Physics B, 2013, 22(3):037202.

[17] Chang C C, Hwang S W, Chen C H, et al. High – efficiency organic electroluminescent device with multiple emitting units. Japanese journal of applied physics, 2004, 43:6418.

[18] Zhang H M, Dai Y F, Ma D G, et al. High efficiency tandem organic light – emitting devices with Al/WO_3/Au interconnecting layer. Applied Physics Letters, 2007, 91(12):123504.

[19] Terai M, Fujita K, Tsutsui T. Capacitance measurement in organic thin – film device with internal charge separation zone. Japanese journal of applied physics, 2005, 44:L1059.

[20] Qi X, Li N, Forrest S R. Analysis of metal – oxide – based charge generation layers used in stacked organic light – emitting diodes. Journal of Applied Physics, 2010, 107(1):014514 – 014514 – 8.

[21] Cheng Y M, Lu H H, Jen T H, et al. Role of the Charge Generation Layer in Tandem Organic Light – Emitting Diodes Investigated by Time – Resolved Electroluminescence Spectroscopy. The Journal of Physical Chemistry C, 2010, 115(2):582.

[22] Meyer J, Kröger M, Hamwi S, et al. Charge generation layers comprising transition metal – oxide/organic interfaces: Electronic structure and charge generation mechanism. Applied Physics Letters, 2010, 96:193302.

[23] Bao Q Y, Yang J P, Tang J X, et al. Interfacial electronic structures of WO_3 – based intermediate connectors in tandem organic light – emitting diodes. Organic Electronics, 2010, 11(9):1578.

[24] Q. Y. Bao, J. P. Yang, Y. Q. Li, et al. Electronic structures of MoO3 – based charge generation layer for tandem organic light – emitting diodes, Appl. Phys. Lett., 97(2010)063303.

[25] Kihyon Hong, Jong – Lam Lee. Charge Generation Mechanism of Metal Oxide Interconnection in Tandem Organic Light Emitting Diodes. The Journal of Physical Chemistry C, 2012, 116(10):6427.

[26] Jin – Peng Yang, Yan Xiao, Yan – Hong Deng, et al. Electric – Field – Assisted Charge

Generation and Separation Process in Transition Metal Oxide - Based Interconnectors for Tandem Organic Light - Emitting Diodes. Advanced Functional Materials,2012,22(3):600

[27] Jens Meyer,Sami Hamwi,Michael Kröger,et al. Transition metal oxides for organic electronics:Energetics,device physics and applications. Advanced Materials,2012,24(40):5408

[28] Kaname Kanai,Kenji Koizumi,Satoru Ouchi,et al. Electronic structure of anode interface with molybdenum oxide buffer layer Org. Electron,10(2009)637

[29] Min Jung Son,Sehun Kima,Soonnam Kwon,et al. Interface electronic structures of organic light - emitting diodes with WO_3 interlayer:A study by photoelectron spectroscopy. Organic Electronics,2009,10(4):637

[30] P. S. Davids,Sh. M. Kogan,I. D. Parker,et al. Charge injection in organic light - emitting diodes:Tunneling into low mobility materials. Applied physics letters,1996,69(15):2270

[31] Davids P S,Campbell I H,Smith D L. Device model for single carrier organic diodes. Journal of Applied Physics,1997,82(12):6319

[32] Campbell A J,Bradley D D C,Lidzey D G. Space - charge limited conduction with traps in poly(phenylene vinylene)light emitting diodes. Journal of applied physics,1997,82(12):6326

[33] Blom P W M,De Jong M J M,Breedijk S. Temperature dependent electron - hole recombination in polymer light - emitting diodes. Applied physics letters,1997,71(7):930-932.

[34] Peng Y Q,Yang Q S,Xing H W,et al. Recombination zone and efficiency in bipolar single layer light - emitting devices:a numerical study. Applied Physics A,2008,93(2):559-564.

[35] Gummel H K. A self - consistent iterative scheme for one - dimensional steady state transistor calculations. Electron Devices,IEEE Transactions on,1964,11(10):455

[36] Scharfetter D L,Gummel H K. Large - signal analysis of a silicon read diode oscillator. Electron Devices,IEEE Transactions on,1969,16(1):64

注:本文曾发表在2014年4月《Journal of semiconductors》第4期上

High efficiency, high energy, CEP – stabilized infrared optical parametric amplifier

Weijun Ling　Xiaotao Geng　Shuyan Guo
Zhiyi Wei　F. Krausz　D. Kim*

摘　要：实验中仅用一块BBO晶体实现了一种高效、可调谐、载波包络相位稳定的近红外光学参量放大器。由CEP稳定激光器产生的白光连续光进行二级放大，采用二类相位匹配实现了34%的转换效率。据我们所知，这是宽带光参量放的国际最高转换效率。这项工作提供了一种简单而有效的载波包络相位稳定的红外高效率参量放大方法。

A high efficiency, tunable, carrier – envelope – phase(CEP) stabilized near – infrared optical parametric amplifier(OPA) is demonstrated with just a single BBO crystal. A white – light continuum produced by a CEP – stabilized laser is seeded into the two stages of a type II OPA system. We achieved a pump – to – signal conversion efficiency of 34% with a single nonlinear crystal. To our knowledge this is the highest conversion efficiencyreported in broadband optical parametric amplification. This work demonstrates a simple and efficient way to produce tunable femtosecond pulses with CEP stabilization.

1. Introduction

Powerful carrier – envelope – phase(CEP) stabilized pulses that are tunable in the visible and/or near infrared spectral range are of special interest due to potential applications to nonlinear optics, ultrafast spectroscopy and attosecond spectroscopy[1,2]. Attosecond spectroscopy requires such pulses for the reproducible generation and precision

---

* 作者简介：令维军(1968—　)，男，甘肃武山人，天水师范学院教授、博士，主要从事激光技术及理论研究。

measurement of isolated attosecond pulses for pump – probe experiments. To this end, pulses in the infrared appear desirable, allowing scaling the photon energy of the attosecond pulse according to the $I\lambda^2$ scaling of the cut – off energy of high – order harmonics, where I is the peak intensity and /lambda is the carrier wavelength [3]. Optical parametric amplification(OPA) represents the most promising approach to achieving this goal.

At present, high energy pulsesare generated mainly by non – collinear optical parametric amplifiers(NOPA) in β – barium borate(BBO) crystals[4-8] or collinear optical parametric amplifier(COPA) in BiB_3O_6(BIBO) crystals[9,10]. There are several ways to generate the broadband radiation needed to seed the OPA for powerful few – cycle pulse production. These include white light continuum(WLC) generation in a single filament in sapphire [4,5], idler wave generation in a noncollinear OPA [6], difference – frequency generation(DFG) with the supercontinuum emerging from b a hollow – core fiber[7,8] or a sapphire plate[9,10]. Such a broadband seed pulse has then been amplified by two or three OPA stages in previous work. For tunable femtosecond OPAs based on BBO, BIBO and LBO crystals overall pump – to – signal conversion efficiencies of about 20% have been reported [11-16]. LBO(as a nonlinear crystal) resulted in the highest pump to signal conversion efficiency of 34% in a 250 mJ, 5 Hz repetition rate OPCPA system pumped by a laser with engineered spatial and temporal beam profiles [17]. However, LBO is not the ideal choice for few – cycle OPA due to its unfavourable gain – bandwidth limit. NOPA implemented with BBO crystals as the nonlinear medium is ideally suited for efficient broadband amplification. A record pump – to – signal conversion efficiency of 32% was recently reported in a broadband optical parametric amplification by a careful optimization of the pump intensity in the crystal, the temporal stretching ratio between pump and seed, and by reusing residual pump energy[18].

CEP stabilization is indispensable for precision attosecond metrology, spectroscopy, and control [1,19-22]. Active and passive technologies have been developed for producing CEP – controlled pulses. The active scheme requires that CEP stabilization be realized by active electronic control loops based on pulse measurements to the pulse phase slips of the oscillator[23-26]. DFG naturally produces pulses with constant CEP(possibly subject to slow drifts, which can be easily stabilized) from any pulse with uncontrolled CEP, resulting in a passive CEP stabilization[27-32]. Both pump and seed are suitable as a non – CEP – stabilized laser for a passive CEP stabilized OPA. The idler beam from white light is often seeded in an optical parametric amplifier Similarly, an OPA driven by a pump and seeded with a signal that are derived from the same (non – CEP – stabilized) radiation yields a (passively) CEP – stabilized idler wave that can be further amplified for powerful CEP – stable pulse generation [34]. Passive CEP stabilization do not require fast feedback loops but generally do rely on active stabilization against slow drifts of the CEP in the amplified output.

CurrentOPA systems require improvements in CEP stabilization, easier tuning and higher efficiency. In this Letter, we report a tunable near – infrared OPA system, with an unprecedented (pump to signal) conversion efficiency of 34%, CEP stabilization and easy tuning. A white – light continuum produced by a CEP stabilized laser is seeded into the two stages of the type II OPA system. We opted for type II phase matching owing to its favourable conversion efficiency[4]. Thanks to the (actively) CEP – stabilized seed, we can avoid any beam degradation that might originate from the idler angle dispersion in the OPA process or the difficulty of separating beams in the DFG process in the passive CEP stabilization schemes. A tunable output requires that the nonlinear crystals in different OPA stages be precisely synchronized and rotated for the same angle and usually mounted on precision rotation mounts with a step motor drive. In order to avoid these difficulties and complications, we have implemented OPA with only one BBO crystal for a two – stage amplification, for an easy and user – friendly synchronous adjustment of the two OPA stages.

A pulse energy of up to 138 μJ at 1350 nm has been obtained at a pump energy of 407 μJ. This corresponds to a pump – to – signal conversion efficiency of 34%, which is (to our knowledge) the highest conversion efficiency in two – stage NOPAefficiencyever reported for broadband optical parametric amplification. Note that we have achieved this efficiency without and shaping of the temporal and spatial profiles of the pump and seed beams. Such shaping affords promise for a substantial further increase of the reported efficiency. The CEP jitter of amplified signals was measured to be 110 mrad by a spectral interferometry based on f – 2f principle, The scheme demonstrates a simple and efficient way to generate tunable femtosecond IR laser pulses with a stabilized CEP.

2. Experimental setup

Figure. 1 shows the experimental layout of the two – stage NOPA setup. A commercial Ti:sapphire multi – pass chirped – pulse – amplification (CPA) laser system (with active CEP stabilization) is used as a pump source to provide 25 fs laser pulses with an energy of 510 μJ at a repetition rate of 3 kHz and a central wavelength of 800 nm. The experimental setup consists of three parts: WLC generation and twoNOPA amplification stages. The driving pulse is split into three parts by two beam splitters(BS). A fraction of the driving pulses(3 μJ) is taken and focused onto a 2 mm thick sapphire plate. An iris placed before the focus lens controls the energy in the sapphire plate and generates a single filament WLC with excellent radial intensity. About 1 μJ of energy is focused into the sapphire plate by a lens with f = 100mm; subsequently, a WLC with excellent beam quality is formed. This is used as the seed for the firststage NOPA(NOPA1). About 100 μJ of energy is split bya beam splitter(with a splitting ratio of 1:5) to pump NOPA1 A 3 mm – thick BBO crystal with an aperture of 17×17 mm^2 is used in NOPA1, which is cutwith a phase matching angle at $\theta = 28°$ $\phi = 30°$ for type II [e(signal) + o(idler) → e(pump)] phase matching. We introduce a 1.7° non – collinear angle (internal) between the pump and the signal. To avoid damaging the parametric crystal or restraining superfluorescence, the beam diameter of the pump laser in the crystal is controlled to be about 1.4 mm by a convex lens(f = 500mm), corresponding to a power intensity of 260 GW/cm^2.

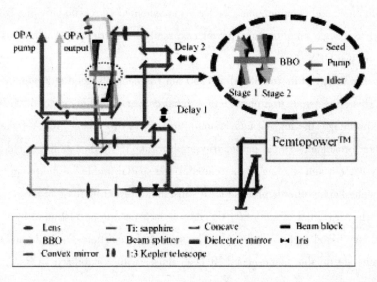

Fig. 1. Schematic of the experimental setup for the generation of hybrid CEP stable pulses based on two – stages NOPA. Inset: Schematic diagram of optical path in a single BBO crystal.

3. Results and Discussion

An amplified signal energy of 3 μJ is obtained in NOPA1. We expand and collimate the seed beam with a Kepler 1:3 telescope to match the beam size of the seed and the pump pulses on the crystal for the second-stage NOPA(NOPA2). Then the collimated signal output from NOPA1 is injected into NOPA2 for further amplification, in which the signal and pump light are incident on the same BBO in the opposite direction and parallel to each other as in NOPA1. The optical path in the BBO is shown by the inset of Fig. 1. About 407 μJ is sent to pump NOPA2 by the second beam splitter. To obtain the best spatial beam quality for the output pulse, the power density on the crystal is optimized by a down-collimator with a ratio of 3:1. We carefully optimize the spatial and temporal match by adjusting the crystal phase matching angle and the time delay for pump pulses in order to obtain a stable amplification in the saturation regime. We measure as much as 138 μJ of amplified signal energy directly after the nonlinear crystal that corresponds to the pump-to-signal conversion efficiency of 34%.

In NOPA2, the tunable signal pulse was amplified and signal spectrum can be tuned from 1100 nm to 1700 nm by changing BBO phase matching angle, (see Fig. 2 (a)). Fig. 2(b) is the conversion efficiency curve corresponding to different signal wavelengths measured by changing phase matching angle. The conversion efficiency varies from 5% around 1100nm and 1700 nm to 34% around 1350 nm.

The OPA output pulse duration is determined by the pump pulse width. The pulse was characterized by spectral phase interferometer for the direct electric field reconstruction(SPIDER) method. The inset in Fig. 2(a) shows the reconstructed pulse in the time domain, which has a full width at half maximum(FWHM) duration of 28 fs around 1350 nm and is comparable to a pump pulse duration of 25 fs FWHM.

The result indicate that the amplified pulse duration does not change relative to the pump light in the IR wavelength range.

It is necessary to restrain the superfluorescent amplified signal(SAS) in order to obtain a good signal beam quality in NOPA2. To do so, the pump intensity for NOPA2 is controlled at 250 GW/cm^2. Simultaneously, we found that the optimization of synchronization between the signal and pump is important to improve beam quality. Figure 3(a) and (b) show the amplified signal beam quality at different times of synchronization. The bad beam quality in Fig. 3(a) is due to imperfect synchronization with a strong SAS. A precise pump and signal synchronization removesmost of the SAS and obtains a perfect beam quality (see Fig. 3(b)).

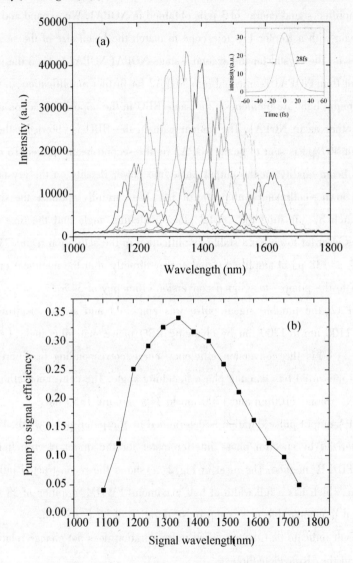

Fig. 2. (a) Tunable output spectrum of the optical parametric amplifier by changing BBO phase matching angle. (b) The pump to signal conversion efficiency curve in the final OPA stage with 407 μJ pump pulses. The highest conversion efficiency is 34% for the signal pulses at 1350 nm.

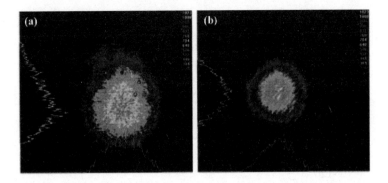

Fig. 3 (a) final amplified pulse profile with SAS; (b) final amplified pulse profile with perfect time synchronization.

In the current tuning range of the spectrum, the relative(with respect to the pump) velocity of the signal wave is opposite to that of the idler wave for type II phase matching in BBO crystal[4]. This means that the energy flow happens mostly in the direction from pump to signal and idler that relaxes crystal length restrictions due to group – velocity(the typical separation length of 2mm for 790nm pumping). This fact results in a higher conversion efficiency. The gain wavelength maintained in a two – stage NOPA (due to the respective parallellity of the pump and signal light) also contributes to a high conversion efficiency. Excellent qualityof pump and signal beams as well as their spatial and temporal matching are also important prerequisites for the high efficiency obtained.

To investigate the effect of NOPA on maintaining a stable CEP in the amplified output, we used an f – 2f interferometer and measured the CEP jitter of the seed as well as the amplified pulse after NOPA2. The f – 2f interferometric spectra were recorded with a time interval of 100 s(see Fig. 4). The analysis of the spectral interference fringes with 50 ms integral time shows that the CEP jitter was 56 mrad(rms) for seed pulse and 110mrad(rms) for amplified signals. Hence, the CEP jitter was increased by approximately a factor of two in the OPA process. A linear Fourier – transform spectral interferometry algorithm analysis showed that the CEP of seed was well maintained through OPA. The results show that the CEP hybrid control method(based on the active control of seed pulse CEP and the passive control of amplified pulse CEP) is feasible and effective for CEP controlled infrared NOPA.

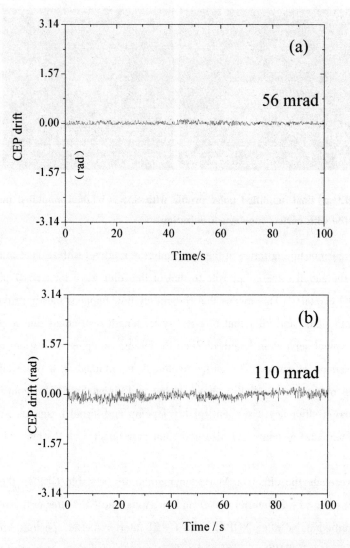

Fig. 4. CEP jitter curve with time interval of 100 s. (a) Seed light CEP jitter is 56 mrad (RMS); (b) the final amplified pulse CEP jitter is 110 mrad (RMS)

4. Conclusion

We developed a highefficiency, high energy, tunable phase – stabilized near – infrared femtosecond OPA at 3 kHz, which consisted of a WLCstage and two non – collinear OPA stages with only one amplification medium (BBO). A hybrid control method stabilized the CEP of the tunable output pulses. The tunable output pulses were still CEP –

stabilizedby an actively CEP – stabilized seed and pump pulse. The measured CEP jitter of theoutput pulses from the OPA was 110 mrad(rms) for an input jitter of 56 mrad. The total tunable spectral range for the signal pulses was from 1100 nm to 1700 nm. The maximum conversion efficiency from pump to signal in the final OPA stage was 34% at 1350 nm. A single BBO crystal demonstrated a high efficiency two – stage NOPA in the IR that can be applicable to various areas of science such as the two – color field experiment for the generation of IAP [35].

This research has been supported by Global Research Laboratory Program [Grant No 2011 – 00131], by Leading Foreign Research Institute Recruitment Program [Grant No 2010 – 00471] andby Max Planck POSTECH/KOREA Research Initiative Program [Grant No 2011 – 0031558] through the National Research Foundation of Korea(NRF) funded by Ministry of Science, ICT & Future PlanningThe work is also supported by the National Natural Science Foundation of China(NO. 60608003, NO. 60878020, and NO. 10874237), and the Instrument Developing Project of the Chinese Academy of Sciences (Grant No. 2010004). Cooperation between the three participating groups has also been greatly stimulated and facilitated by the Max – Planck – Center for Attosecond Science funded by the Max Planck Society.

References

[1] F. Krausz and M. Ivanov, "Attosecond physics," Rev. Mod. Phy. 81,163(2009).

[2] Hentschel et al, "Attosecond metrology", Nature 414, p. 509(2001)

[3] C. Hernández – García,, J. A. Pérez – Hernández, T. Popmintchev, M. M. Murnane, H. C. Kapteyn, A. Jaron – Becker, A. Becker, and L. Plaja, "Zeptosecond High Harmonic keV X – Ray Waveforms Driven by Midinfrared Laser Pulses," Phy. Rev. Lett. 111,033002(2013).

[4] Kent R. Wilson and Vladislav V. Yakovlev, "Ultrafast rainbow: tunable ultrashort pulses from a solid – state kilohertz system," J. Opt. Soc. Am. B 14,444 – 448(1997).

[5] G. Cerullo and S. De Silvestri, "Ultrafast optical parametric amplifiers," Rev. Sci. Instrum. 74,1 – 18(2003).

[6] Chunmei Zhang, Pengfei Wei, Yansui Huang, Yuxin Leng, Yinghui Zheng, Zhinan Zeng, Ruxin Li, and Zhizhan Xu, "Tunable phase – stabilized infrared optical parametric amplifier for high – order harmonic generation," Opt. Lett. 34,2730 – 2732(2009).

[7] C. Vozzi, G. Cirmi, C. Manzoni, E. Benedetti, F. Calegari, G. Sansone, S. Stagira, O. Svelto, S. De Silvestri, M. Nisoli, and G. Cerullo, "High – energy, few – optical – cycle pulses at 1.5 μm with passive carrier – envelope phase stabilization," Opt. Express 14,10109 – 10116(2006).

[8] C. Vozzi, F. Calegari, E. Benedetti, S. Gasilov, G. Sansone, G. Cerullo, M. Nisoli, S. De Silvestri, and S. Stagira, "Millijoule – level phase – stabilized few – optical – cycle infrared parametric source," Opt. Lett. 32, 2957 – 2959(2007).

[9] F. Silva, P. K. Bates, A. Esteban – Martin, M. Ebrahim – Zadeh, and J. Biegert, "High – average – power, carrier – envelope phase – stable, few – cycle pulses at 2.1μm from a collinear BiB3O6 optical parametric amplifier," Opt. Lett. 37, 933 – 935(2012).

[10] M. Ghotbi, M. Beutler, V. Petrov, A. Gaydardzhiev, and F. Noack, "High – energy, sub – 30 fs near – IR pulses from a broadband optical parametric amplifier based on collinear interaction in BiB_3O_6," Opt. Lett. 34, 689 – 691(2009).

[11] Z. Sun, M. Ghotbi, and M. Ebrahim – Zadeh, "Widely tunable picosecond optical parametric generation and amplification in BiB3O6," Opt. Express 15, 4139 – 4148(2007).

[12] O. V. Chekhlov, J. L. Collier, I. N. Ross, P. K. Bates, M. Notley, C. Hernandez – Gomez, W. Shaikh, C. N. Danson, D. Neely, P. Matousek, S. Hancock, and L. Cardoso, "35 J broadband femtosecond optical parametric chirped pulse amplification system," Opt. Lett. 31, 3665 – 3667 (2006).

[13] M. Ghotbi, M. Ebrahim – Zadeh, V. Petrov, P. Tzankov, and F. Noack, "Efficient 1kHz femtosecond optical parametric amplification in BiB3O6 pumped at 800 nm," Opt. Express 14, 10621 – 10626(2006).

[14] Zhang J, Zhang Q L, Zhang D X, Feng B H and Zhang J Y, "Generation and optical parametric amplification of picosecond supercontinuum", Applied Optics, Vol. 49, Issue 34, pp. 6645 – 6650(2010).

[15] M. Ghotbi, M. Beutler, V. Petrov, A. Gaydardzhiev, and F. Noack, "High – energy, sub – 30 fs near – IR pulses from a broadband optical parametric amplifier based on collinear interaction in BiB_3O_6", Opt. Lett. 34, 689(2009).

[16] L. J. Waxer, V. Bagnoud, I. A. Begishev, M. J. Guardalben, J. Puth, and J. D. Zuegel, "High – conversion – efficiency optical parametric chirped – pulse amplification system using spatiotemporally shaped pump pulses," Opt. Lett. 28, 1245 – 1247(2003).

[17] V. Bagnoud, I. A. Begishev, M. J. Guardalben, J. Puth, and J. D. Zuegel, "5 Hz, > 250 mJ optical parametric chirped – pulse amplifier at 1053 nm," Opt. Lett. 30, 1843(2005).

[18] Hanieh Fattahi, Christoph Skrobol, Moritz Ueffing, Yunpei Deng, Alexander Schwarz, Yuichiro Kida, Vladimir Pervak, Thomas Metzger, Zsuzsanna Major, and Ferenc Krausz, "High efficiency, multi – mJ, sub 10 fs, optical parametric amplifier at 3 kHz," CLEO Technical Digest OSA 2012.

[19] Kienberger R, Goulielmakis E, Uiberacker M, et al., "Atomic transient recorder," Nature 427, 817 – 828(2004).

[20] Sansone G, Benedetti E, Calegari F, et al., "Isolated single – cycle attosecond pulses,"

Science 314,443 - 446(2006).

[21] M Schultze, E Goulielmakis, M Uiberacker, M Hofstetter, J Kim, D Kim, F Krausz and U Kleineberg, "Powerful 170 - attosecond XUV pulses generated with few - cycle laser pulses and broadband multilayer optics," New Journal of Physics 9,243(2007).

[22] S. Zherebtsov, Th. Fennel, J. Plenge, E. Antonsson, I. Znakovskaya, A. Wirth, O. Herrwerth, F. Süßmann, P. Peltz, I. Ahmad, S. A. Trushin, V. Pervak, S. Karsch, M. J. J. Vrakking, G. Graf, M. I. Stockman, F. Krausz, E. Rühl, and M. F. Kling, "Controlled near - field enhanced electron acceleration from dielectric nanospheres with intense few - cycle laser fields," Nature Physics 7,656(2011).

[23] Cundiff S. T. , "Phase stabilization of ultrashort optical pulses," Phys. D: Appl. Phys. 35,43 - 59(2002).

[24] Chang Z H, "Carrier - envelope phase shift caused by grating - based stretchers and compressors," Appl. Opt 45,8350 - 8353(2006).

[25] Gagnon E, Thomann I, Paul A, et al. , "Long - term carrier - envelope phase stability from a grating - based, chirped pulse amplifier," Opt. let 31,1866 - 1868(2006).

[26] Li C Q, Moon E, and Chang Z H, "Carrier - envelope phase shift caused by variation of grating separation," Opt. Lett. 31,3113 - 3115(2006).

[27] Fuji T, Apolonski A, Krausz F, "Self - stabilization of carrier - envelope offset phase by use of difference - frequency generation," Opt. Lett. 29,632 - 634(2009).

[28] Manzoni C, Vozzi C, Benedetti E, et al. , "Generation of high - energy self - phase - stabilized pulses by difference - frequency generation followed by optical parametric amplification," Opt. Lett. 31,963 - 965(2006)

[29] Manzoni C, Cerullo G, Silvestri S De, "Ultrabroadband self - phase - stabilized pulses by difference - frequency generation," Opt. Lett. 29,2668 - 2670(2004).

[30] G. Cirmi, C. Manzoni, D. Brida, S. De Silvestri, and G. Cerullo, "Carrier - envelope phase stable, few - optical - cycle pulses tunable from visible to near IR," J. Opt. Soc. Am. B 25, B62 - B69(2008).

[31] A. Baltuška, T. Fuji, and T. Kobayashi, "Controlling the carrier – envelope phase of ultrashort light pulses with optical parametric amplifiers," Phys. Rev. Lett. 88,133901(2002).

[32] C. Manzoni, D. Polli, G. Cirmi, D. Brida, S. De Silvestri, and G. Cerullo, "Tunable few - optical - cycle pulses with passive carrier - envelope phase stabilization from an optical parametric amplifier," Appl. Phys. Lett. 90,171111(2007).

[33] E. Moon, C. Li, Z. Duan, J. Tackett, K. L. Corwin, B. R. Washburn, and Z. Chang, "Reduction of fast carrier - envelope phase jitter in femtosecond laser amplifiers," Opt. Express 14, 9758 - 9763(2006).

[34] C. Manzoni, D. Polli, G. Cirmi, D. Brida, S. De Silvestri, and G. Cerullo, "Tunable few -

optical – cycle pulses with passive carrier – envelope phase stabilization from an optical parametric amplifier," Appl. Phys. Lett. 90,171111(2007).

[35]Byunghoon Kim,Jungkwen Ahn,Yongli Yu,Ya Cheng,Zhizhan Xu,and Dong Eon Kim, "Optimization of multi – cycle two – color laser fields for the generation of an isolated attosecond pulse",Optics Express 16,10331(2008)

注:本文曾发表在2014年《Journalof Optics》第19期上

1.91μm passively continuous-wave mode-locked Tm:LiLuF₄ laser

WeiJun Ling   Tao Xia   Zhong Dong   LiangFang You
YinYan Zuo   Ke Li   QinLiu   FeiPing Lu   XiaoLong Zhao*

摘 要：我们首次在国际上实现了 TM:LiluF4 激光器连续锁模运转。利用砷化镓半导体可饱和吸收镜作为锁模元件,实现了 1914 纳米波长下脉冲小于 14 皮秒的锁模运转,对应最大输出功率为 200 毫瓦,同时实验研究了输出功率对泵浦参数的依赖关系。

We report, to our knowledge, the first demonstration of continuous - wavemode - locking in a Tm:LiLuF₄ laser. By using a GaAs - based semiconductor saturable absorber mirrors(SESAMs), we achieved the mode - locked operation with pulses as short as 14 ps at 1914nm. The maximum output power is 200 mW. The dependence of the operational parameters on the pump power has been investigated experimentally.

1. Introduction

Tm - ,Ho - and Tm,Ho - doped mode - locked bulk lasers at 2 μm are attracted much more attentions due to the emission wavelengths near the absorption peak of water and in the atmospheric window band, which has important applications in surgical operation[1], eye - safe lidar[2], electro - optical countermeasure[3], and remote sensing[4]. In recent years, the ultrafast lasers at 2 μm became also the potential laser sources for chirped - pulse amplification[5], synchronously pumped optical parametric oscillator[6], mid - IR frequency comb [7], supercontinuum generation [5] and material

---

* 作者简介:令维军(1968—),男,甘肃武山人,天水师范学院教授、博士,主要从事激光技术及理论研究。

processing[8].

Up to now, the passive mode – locking technology is the main method to obtain ultrafast pulses 2 μm. In 2009, Cho W B et al. demonstrated a stable and self – starting mode – locked Tm:KLu(WO$_4$)$_2$ laser using a transmission – type single – walled carbon nanotube(SWCNT) as saturable absorber(SA)[9]. Since then, passive mode locking operations were obtained for a variety of Tm – doped and Tm, Ho – co – doped gain media with different mode – locking devices, such as SESAM, SWCNT, graphene and 2D material(multilayer MoS$_2$, WS$_2$ ect.) and so on. Meanwhile, the novel SAs such as SWCNT[10-11], graphene[12] and 2D material[13] have been explored as passive mode – locking devices. For example, With a graphene saturable absorber, 86 – fs pulses was demonstrated from a Tm:MgWO$_4$ laser in 2017[14]. However, these SAs have relatively high mode – locking threshold because of high loss. With the development of semiconductor technology, the operating wavelength of SESAMs has been extended to infrared range and become one of the most popular devices in 2 μm mode – locked lasers[15] due to its low loss.

As the most widely used mode – lockingdevice, SESAMs were successfully used in Tm – doped and Tm, Ho – co – doped gain media lasers to achieved mode – locking operation. Using tungstate materials as gain media, Lagatsky A A et al. realized ~100 fs mode – locked operations from Tm, Ho:NaY(WO$_4$)$_2$ and Tm:KYW oscillators[16-17]. Gluth A et al. obtained 3 ps pulses from a Tm:YAG laser[18] and Kong LC et al realized a mode – locked Tm:CaYAlO$_4$ laser with 30 ps pulses[19]. Later, Wang Y et al. demonstrated a Tm:CaYAlO$_4$ oscillator with the pulse duration of 650 fs[20]. In 2017, C Luan et al. realized a mode – locked Tm:LuAG laser with the pulse duration of 13.6 ps[21]. Moreover, Tm – doped cubic sesquioxides are also very successful gain materials. In 2012, Lagatsky A A et al obtained ~100 fs pulses from Tm:Sc$_2$O$_3$ and Tm:Lu$_2$O$_3$ ceramic lasers, respectively[22-23]. In addition, Ma J et al demonstrated a laser diode pumped Tm:CLNGG laser with output of 479 fs pulses[24]. The above mentioned mode – locked lasers have always smooth and non – modulated spectra and the center wavelength is longer than the water absorption peak of 1.95μm, which can avoid Q – switched instability due to the strong absorption of water in the air[25]. Because of the spectral modulation and the wavelength close to the absorption peak of water, it is hard to achieve continuous – wave mode – locked operation from Tm – doped and Tm, Ho –

co-doped fluoride media. Up to now, the only mode-locked operation was obtained from a Tm:GdLiF4 laser[26].

Compared with other Tm-doped crystal, Tm:LiLuF$_4$ crystal characterized by low phonon energy, big absorption cross section and low absorption for laser, was suitable for low threshold operation[27]. In 2007, Cornacchia F et al. investigated the laser performance of Tm:LiLuF$_4$ and obtained the output power of 280 mW with the spectra covering of 1985-2038nm[28]. In 2009, Xiong J et al achieved the maximum CW power of 7.16 W with the central wavelength of 1.92 μm[29]. Recently, Zou X and Zhang X et al demonstrated a Q-switched mode-locked Tm:LiLuF$_4$ laser with MoS$_2$ and Cr:ZnS as saturable absorber, respectively[30-31], however, the continuous-wave mode-locked operation of Tm:LiLuF$_4$ has not been realized. In this paper, to our knowledge, we demonstrated a stable passively continuous-wave mode-locked Tm:LiLuF$_4$ laser for the first time. A stable and self-starting passive continuous mode-locking operation is achieved by using a SESAM from a Tm:LiLuF$_4$ laser with output power of 200 mW and pulse duration of 14 ps at the central wavelength of 1914 nm and a repetition rate of 100 MHz.

2. Experimental setup

The experimental setup is shown in Fig. 1. A typical X-shape cavity was used in our experiment.

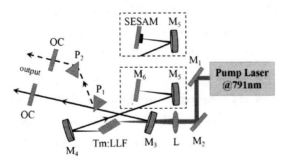

Fig. 1. (color online) Experimental setup of the passively mode-locked Tm:LiLuF$_4$ laser

The pump source is a Ti:sapphire laser with a wavelength range of 740-844nm, and the output wavelength can be tuned to the 791nm wavelength by a birefringent filter. L is a focusing lens with a focal length of 120 mm, which is antireflection(AR) - coated at 791 nm. A Brewster - cut 2 at. % doped Tm:LiLuF$_4$ crystal with dimensions of $3 \times 3 \times 12$ mm^3 is used as the gain medium, which is wrapped with indium foil and tightly mounted on a water - cooled copper heat sink at the temperature of 13 ℃ to remove the heat deposited inside of crystal. M$_1$ and M$_2$ are plane mirrors, which are highly reflected(>99.9%) at the pump wavelength. M$_3$ and M$_4$ are dichroic mirrors with radii of curvature(ROC) of 100 mm, which have high reflection in the range of 1800 - 2075 nm(>99.9%) and high transmission in the range of 770 - 1050 nm(>95%). M$_5$ is a concave mirror with the ROC of 100 mm, which is highly reflected in the range of 1800 - 2075 nm(>99.9%). M$_6$ is a plane mirror with reflectivity >99.9% at the wavelength range of 1800 - 2075 nm. OCs are the output couplers, which have transmissions of 1.5%, 3% and 5% in the experiment. A SESAM on GaAs substrate with the non - saturable loss of 2% and the relation time of 10 ps is used to start mode locking. The damage threshold of the SESAM is 4 mJ/cm^2. A pair of CaF$_2$ prism(P$_1$ and P$_2$) with a tip - to - tip distance of 300 mm is used for intracavity dispersion compensation.

Based on with ABCD matrix theory, the calculated laser mode size in the crystal is 50 μm, and the radius of the pump spot in the gain medium is set to 24 μm, which is optimized for realization of the efficient fundamental mode output[32]. Meanwhile, the extreme small laser mode isadvantageous for laser operation with low threshold.

3. Analysis and discussion of experimental results

In experiment, the laser threshold is as low as 80 mW(absorbed pump power). To the best of our knowledge, this is the lowest threshold in Tm - doped fluoride lasers. The absorption efficiency in lasing and non - lasing conditions for OCs with different transmissions are measured and compared to investigate the absorption and emission characters of Tm:LiLiF$_4$ crystal, as shown in Fig. 2(a). When the oscillator operated in CW or mode - locking operations, the absorption efficiency is about 65.5%. It shows that the transmission of output couplers had little influence on the absorption efficiency of the gain medium. When the lasing condition is blocked, the absorption efficiency is reduced to 34.3%, which is much lower than the laser running. This is because thestimulated radiation will consume a large number of upper level particles in lasing condition, the absorption rate of crystal is higher. when non laser operation, only spontaneous radiation will consume a few upper level particles, which reduces the efficiency of crystal absorption.

large number of upper level populations is consumed by the laser when it is running. In the state of laser operation, in addition to spontaneous radiation, ;

The output performance of the Tm:LiLuF$_4$ laser is shown in Fig. 2(b). The output coupler with high reflection is used for the alignment of the cavity. The threshold is only 80 mW(absorbed pump power) when a 1.5% output coupler is used, and the slop efficiency is about 26.8% In this case, the maximum output power of 632 mW is achieved with aoptical-to-optical efficiency of 17.1%. When a 3% output coupler is used, the threshold increases to 112 mW and the slop efficiency increases to 33%. The maximum output power is 741 mW, corresponding to an optical-to-optical efficiency of 20%, and when the absorbed pump power is higher than 2.33W, the laser tends to saturation. With even higher transmission of 5% as the output coupler, 766 mW output is achieved and the optical-to-optical efficiency is 20.8%, and when the absorbed pump power is higher than 2.31W, the laser tends to saturation. The laser threshold is 131 mW and the slop efficiency is 34.9%.

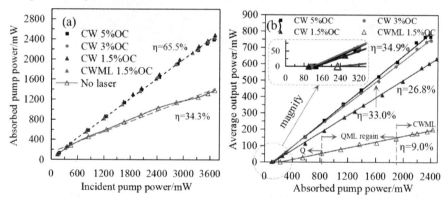

Fig. 2. (color online) (a) Absorption efficiency of Tm:LiLuF$_4$. (b) Average output power versus absorbed pump power in CW(black, green and blue line) and mode locking(red line).

The mirror of the M$_6$ mirror is replaced by a SESAM to achieve the mode-locking operation. To improve the energy density on the surface of the SESAM, a concave mirror with a radius of 100 mm is used. The stable mode-locking operation is realized by finely adjusting the position of SESAM. The laser mode size on the SESAM is calculated to be about 140 μm, corresponding to the energy fluence about 336 μJ/cm^2. It is much larger than the saturable fluence (70 μJ/cm^2) of SESAM. A pair of CaF$_2$ prisms is inserted into the cavity to compensate the intracavity dispersion and shorten the pulse duration. With a 1.5% output coupler, the cavity begins output with Q – switched pulse trains when the absorbed pump power is 228 mW. Further increase the the absorbed pump power to 822 mW, the Q – switched mode – locking operation is achieved. The stable mode – locking operation is realized at the absorbed pump power of 1.9 W. The maximum output power of 200 mW is achieved at 2.4 W absorbed pump power, corresponding to an optical – to – optical efficiency of 5.4%.

Fig. 3. Mode – locked pulse train in the time scales of 1 ms/div and 10 ns/div

The mode – locked pulse train is measured with a high – speed photodiode (PD) and a 200 – MHz digital oscilloscope (RIGOL, DS4024), as shown in Figure 3. It shows the mode – locked pulse train has a repetition rate of 100 MHz, corresponding to a maximum pulse energy of 2 nJ.

The spectra of the mode – locked pulses are measured with a commercial spectrometer (AvaSpec – NIR256 – 2.5TEC). The full width at half maximum (FWHM) bandwidth is 16 nm at central wavelength of 1914 nm, as shown in Fig. 4. The mode – locked wavelength previously reported is more than 2μm, which is away from the water molecule absorption peak of 1.93μm and relatively easy to implement locking. In the experiment, a dehumidifier was used to reduce the humidity of the laboratory to 17%, greatly reducing Q modulation of the water molecules in air and obtained mode – locked operation near the water absorption peak. The pulse duration of the pulses are estimated with intensity autocorrelation traces (APE, pulse check 50). The FWHM width of the autocorrelation trace is 20 ps and the pulse duration is 14 ps with an assumption of Gaussian pulse shape, corresponding to the time – bandwidth product of 18.3 (Gauss – shape pulse, ideal 0.441), which indicates the output pulses have a strong chirp.

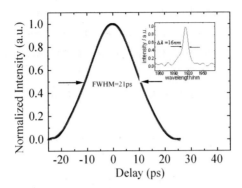

Fig. 4. measured autocorrelation trace (black line) and Mode – locked spectrum (inset)

4. Conclusion and perspectives

In conclusion, we demonstrated, to our knowledge, a continuous and self – starting passively mode – locked Tm:$LiLuF_4$ laser at 1914 nm with a SESAM as mode – locked device for the first time. At an absorbed pump power of 2.43 W, the maximum average output power was 200 mW with a repetition rate of 100 MHz and a typical pulse duration of 14 ps. The spectral bandwidth of 16 nm is theoretically supported of 336 fs pulses. Further optimization of modulation depth and cavity dispersion is expected to achieve a much shorter mode – locked pulse. Such kind of mode – locked pulses in mid – infrared spectral range will open wide applications such as ultrafast molecule spectroscopy, mid – IR pulse generation, laser radar, atmospheric environment monitoring, etc.

Acknowledgment

This work was supported by the National ScienceFoundation of China (Grant No. 11774257, 61465012, 61564008, 61461046, 61665010, 61661046), theNational key research and development program (2017YFB045200) and Tianshui science and technology support program project (Grant Nos. 2018 – FZJHK – 2704).

References

[1] Hüttmann G, Yao C, Endl E 2005 *Medical Laser Application* **20** 135

[2] Nobuo S, Newton S, Kinpui C and Dennis K K 1990 *Opt. Lett.* **15** 302

[3] Lin Y F, Zhang G, Zhu H Y, Huang C H and Wei Y 2008 *Journal of Russian Laser Research* **29** 520

[4] Kaufman Y J, Kleidman R G, Hall D K, Martins J V and Barton J S 2002 *Geophysical Research Letters* **29** 28 - 1

[5] Hong KH, Lai CJ, Siqueira JP, Krogen P, Moses J, Chang CL, Stein GJ, Zapata LE, Kärtner FX 2014 *Opt. Lett.* **39** 3145

[6] Sun J H, Gale B J S, Reid D T 2007 *Opt. Lett.* **32** 1414

[7] Khodabakhsh A, Ramaiah – Badarla V, Rutkowski L, Johansson A C, Lee K F, Jiang J, Mohr C, ermann M E F, Foltynowicz A 2016 *Optics Letters* **41** 2541

[8] Yao J, Zhang B, Yin K, Yang L, Hou J, Lu Q 2016 *Optics Express* **24** 15093

[9] Voisiat B, Gaponov D, Gecys P, Lavoute L, Silva M, Hideur A, Ducros N, Raciukaitis G 2015 *Proc. SPIE* **9350** 935014

[10] Cho W B, Schmidt A, Yim J H, Choi S Y, Lee S, Rotermund F, Griebner U, Steinmeyer G, Petrov V, Mateos X, Pujol M C, Carvajal J J, Aguiló M, Díaz F 2009 *Opt. Lett. express* **17** 11007

[11] Schmidt A, Young C S, Dong – Il Y, Rotermund F, Mateos X, Segura M, Diaz F, Petrov V, Uwe G 2012 *Applied Physics Express* **5** 2704

[12] Schmidt A A, Koopmann P P, Huber G G, Peter F, Sun Y C, Dong I Y, Fabian R, Valentin P, Uwe G 2012 *Lett. express* **20** 5313

[13] Wan H, Cai W, Wang F, Jiang S Z, Xu S C, Liu J 2016 *Optical and Quantum Electronics* **48** 11

[14] Wang Y, Chen W, Mero M, Zhang L, Lin H, Lin Z, Zhang G, Rotermund F, Cho YJ, Loiko P, Mateos X, Griebner U, Petrov V 2017 *Opt. Lett.* **42**, 3076

[15] Hou J, Zhang B, Su X, Zhao R, Wang Z, Lou F, He J 2015 *Optics Communications* **347** 88

[16] Lagatsky A A, Han X, Serrano M D, Cascales C, Zaldo C, Calvez S, Dawson M D, Gupta J A, Brown C T A, Sibbett W 2010 *Opt. Lett.* **35** 3027

[17] Lagatsky A A, Calvez S, Gupta J A, Kisel V E, Kuleshov N V, Brown C T A, Dawson M D, Sibbett W 2011 *Lett. express* **19** 9995

[18] Gluth A, Wang Y, Petrov V, Paajaste J, Suomalainen S, Härkönen A, Guina M, Steinmeyer G, Mateos X, Veronesi S, Tonelli M, Li J, Pan Y, Guo J, Griebner U 2015 *Lett. express* **23** 1361

[19] Kong L C, Qin Z P, Xie G Q, Xu X D, Xu J, Yuan P, Qian L J 2015 *Opt. Lett.* **40** 356

[20] Wang Y C, Xie G Q, Xu X D, Di J Q, Qin Z P, Suomalainen S, Guina M, Härkönen A, Agnesi A, Griebner U, Mateos X, Loiko P and Petrov V 2016 *Optical Materials Express* **6** 131

[21] Luan C, Yang K, Zhao J, Zhao S, Li T, Zhang H, He J, Song L, Dekorsy T, Guina M, Zheng L 2017 *Opt. Lett.* **42** 839

[22] Lagatsky A A, Koopmann P, Fuhrberg P, Huber G, Brown C T A, Ibbett W 2012 *Opt. Lett.* **37** 437

[23] Lagatsky A A, Antipov O L and Sibbett W 2012 *Opt. express* **20** 19349

[24] Ma J, Xie G Q, Gao W L, Yuan P, Qian L J, Yu H H, Zhang H J, Wang J Y 2012 *Opt. Lett.* **37** 1376

[25] Zhang HN, Chen XH, Wang QP, Zhang XY, Chang J, Gao L, Shen HB, Cong ZH, Liu ZJ, Tao XT, Li P 2013 *Opt. Lett.* **38** 3075

[26] Coluccelli N, Galzerano G, Gatti D, Lieto A D, Tonelli M, Laporta P 2010 *Applied Physics B* **101** 75

[27] Peng H, Zhang K, Zhang L, Hang Y, Xu J Q, Tang Y L, Cheng Y, Xiong J, Zhao C C, Chen G Z, He X M 2010 *Chinese Optics Letters* **8** 63

[28] Cornacchia F, Parisi D, Tonelli M 2008 *IEEE Journal of Quantum Electronics* **44** 1076

[29] Xiong J, Peng H Y, Zhao C C, Hang Y, Zhang L H, He M Z, He X M, Chen G Z 2009 *Laser Physics Letters* **6** 868

[30] Zou X, Leng Y X, Li Y Y, Feng Y Y, Zhang P X, Hang Y, Wang J 2015 *Chinese Opt. Lett.* **3** 70

[31] Zou Y W, Wang Q, Zhang Y D, Zhang L 2011 *Chinese Journal of Lasers* **38** 0402004

[32] Silvestri S D, Laporta P and Magni 1986 *Opt. Lett.* **11** 785

注:本文曾发表在2018年《Optics and Laser Technology》第2期刊上

Electronic Structures of Anatase $(TiO_2)_{1-x}(TaON)_x$ Solid Solutions: First-Principles Study

Wenqiang Dang   Hungru Chen   Naoto Umezawa   Junying Zhang*

摘 要：对宽带隙光功能材料的可见光敏化，是使之高效利用太阳能的重要课题。尽管氮掺杂锐钛矿相二氧化钛的研究工作已经被广泛研究，但由于用氮替代氧是异价掺杂，这会促使第二相的生成或者成为缺陷，障碍光生载流子的迁移。本文用第一性原理计算了$(TiO_2)1-x(TaON)_x(0 \leqslant x \leqslant 1)$固溶体的电子结构。计算结果表明，固溶体的带隙比$TiO_2$窄得多，而禁带隙中没有引入任何局域能量状态。另外，与纯TiO_2相比，当TaON含量超过0.25时，固溶体具有直接带隙，有利于光吸收。固溶体的价带顶(VBM)比由O_2p非键态组成的锐钛矿型TiO_2的价带顶具有更高的能量，它主要由N_2p态与O_2p、Ti_3D或Ta_5d轨道杂化而成。另一方面，在TiO_2中引入TaON会因π相互作用形成d-d成键态，并且由于某些金属原子之间的距离缩短，从而大大降低了导带底(CBM)。因此，锐钛矿型$(TiO_2)_{1-x}(Taon)_x$有望成为一种很有前途的可见光吸收体。此外，还发现一些原子构型具有非常窄的带隙。

Sensitizing wide band gap photo-functional materialsunder visible-light irradiation is an important task for efficient solarenergy conversion. Although nitrogen doping into anatase TiO_2 has been extensively studied for this purpose, it is hard toincreasenitrogen content in anatase TiO_2 becauseof the aliovalent nitrogen substituted for oxygen, leading to the formationof secondary phases or defects that hampers the migration of photoexcited charge carriers. In this paper, electronicstructures of $(TiO_2)_{1-x}(TaON)_x$

---

* 作者简介：党文强(1983—   )，男，甘肃秦安人，天水师范学院讲师、博士，主要从事计算物理研究。

($0 \leqslant x \leqslant 1$) solid solutions, in which the stoichiometry is satisfied with theco – substitution of Ti forTa along with O for N, are investigated within the anatase crystal structure using first – principles calculations. Ourcomputational results show that the solid solutions have substantially narrower band gaps than TiO_2, without introducingany localized energy statesin the forbidden gap. In addition, in comparison with the pristine TiO_2, the solid solution has adirect band gap when the content of TaON exceeds 0.25, which is advantageous fo r light absorption. The valence bandmaximum(VBM) of the solid solutions, which is mainly composed of N 2p states hybridized with O 2p, Ti 3d or Ta 5d orbitals, is higher in energy than that of pristine anatase TiO_2 consisting of non – bonding O 2p states. On the other hand, incorporatingTaON into TiO_2 causes the formation of d – d bonding states through π interactions and substantially lowers the conductionband minimum(CBM) because of the shortened distance between some metal atoms. As a result, the anatase $(TiO_2)_{1-x}(TaON)_x$ is expected to become a promising visible – light absorber. In addition, some atomic configurations are found to possessexceptionally narrow band gaps.

1. Introduction

Since the discovery of Fujishima and Honda effect in 1972,[1] photocatalysis has drawn extensive attention both in the fieldsof environmental remediation and energy conversion. Titaniumdioxide(TiO_2), the most well – known photocatalyst, is active only-under ultraviolet(UV) irradiation, and exhibits poorperformance under the shine of visible light which accounts forthe majority of sunlight. One effective way to enhance thevisible – light photocatalytic activity is doping nonmetal ions(such as C, N, B, S and F),2 – 9 metal ions10, 11 or both12 – 15 into theTiO_2 lattice to introduce extra energy states in the band gap. Ingeneral, the doping concentration especially that of the cationsis less than 10 at%, because the dopants usually has limitedsolubility in the TiO_2 matrix and high dopant concentrationdeduces the formation of secondary compound. Replacingoxygen by nitrogen in TiO_2 could substantially shift the photoabsorption edge, leading to the visible – lightphotocatalyticactivity.[2,7-9,16] However, it was reported that when Nconcentration exceeds over the solubility limit, nitrogen atomsstart to occupy interstitial sites, leading to transformation fromthe active anatase to less – active rutile phase.[17] Heavily N – dopedanatase TiO_2 up to 15% was grown by a novel epitaxialmethod,[18] although it is still challenging to achieve such a highconcentration of nitrogen in a powder formed sample forphotocatalysis application. As a result, the band structure wasunable to be tuned in a large scale. In addition, in such dopedTiO$_2$ photocatalysts, it's the dopants that cause visible – lightabsorption, thus the absorption coefficient depends on theconcentration of dopants. As a consequence, only a smallmodification of the photoabsorption edge can be expected bythe doping scheme. Furthermore, the impurity levels created bydopants in the forbidden band of the material are usuallydiscrete, which would appear disadvantageous for themigration of photogenerated holes. In contrast, solid solutioncontains N as constituent element that forms the top of thevalence band. Thus, photogenerated holes can migratesmoothly in the valence band of the material, which isparticularly advantageous for water oxidation involving 4 – electron transfer.[19] Up to now, the maximum solubility has beenobtained in the TiO$_2$ – ZrO$_2$ system,[20,21] incorporating up to 30 at% ZrO$_2$ into TiO$_2$, which enhanced the photocatalytic activity underUV light irradiation but failed under visible light because ZrO$_2$ has larger band gap than TiO$_2$.

High visible light photocatalytic activity has been realized inother systems than TiO$_2$ by forming solid solution using twomaterials with a similar crystal structure. For example, a newtype of oxynitride with a unique composition and structure wasobtained by compositing GaN and ZnO, both possessing wideband gap with a wurtzite – type structure, which- achieved watersplitting into H$_2$ and O$_2$ under visible light irradiation.[22,23] A novelseries of perovskite – type solid solution photocatalysts AgNbO$_3$ – SrTiO$_3$ were powerful for oxidizing H$_2$O into O$_2$ from aqueousAgNO3 solution under visible light. 24 Solid solution β – AgAl$_{0.6}$Ga$_{0.4}$O exhibited 35 and 63 times higher photocatalyticactivities than two terminus materials β – AgAlO$_2$ and β – AgGaO$_{2.25}$ Solid solution(AgIn)xZn$_{2(1-x)}$S$_2$ is an activephotocatalyst for H$_2$ evolution under visible – light irradiationeven though AgInS$_2$ and ZnS hardly possesses any activity undervisible – light irradiation.[26]

δ – TaON phase(anatase – type structure) tantalum oxide nitridepowder was successfully prepared by reaction of gaseousammonia with an amorphous tantalum oxide precursor.[27] AnataseTaON film with band gap of 2.37 eV was obtained byusing nitrogen plasma assisted pulsed laser deposition. 28 J. Grinset al. prepared(TiO$_2$)$_{1-x}$(TaON)$_x$ (0.52 < x < 0.87) solid solution withanatase structure by ammonolysis of Ti – Ta gels.[29] These reportssuggest that(TiO$_2$)$_{1-x}$(TaON)$_x$ has a potential to be used as avisible – light photocatalytic solid solution. Herein, we studied thecrystal and electronic structures of(TiO$_2$)$_{1-x}$(TaON)$_x$ solidsolutions(x = 0.25, 0.5, 0.75) using first – principles calculations. The band gaps of the solid – solution narrow in comparison withthat of TiO_2, but is not a simple function of x. Increasing TaONcontent tunes the solid solution from indirect band gap to directband gap, without forming any localized energy bands. This isdifferent with the N and Ta co-doped TiO_2 by R. Long et al.[14] They found that incorporation of Ta and N into TiO_2 leads to theformation of continuum-like fully occupied N 2p-Ta 5d hybridized states above the top of the valence band as well asTa 5d orbitals located at the bottom of the conduction band. Asfor anataseTaON whose photocatalytic activity has not beenevaluated experimentally, we theoretically determined the CBMand VBM positions, demonstrating its great potential as avisible-light responsive photocatalyst, in agreement with theprevious study of T. Lüdtke et al., indicating that δ-TaON is apromising photocatalytic material for water splitting.[27]

2. Calculation methods

The density functional theory (DFT) calculations were performedusing the spin-polarized projector augmented wave (PAW) pseudopotentials[30] via the Vienna Ab initio Simulation Program (VASP).[31] The exchange and correlation energy was treatedusing the generalized gradient approximation (GGA) viaPerdew-Burke-Ernzerhof (PBE) prescription.[32] The constitutedatomic valence states adopted were N $2s^{22}p^3$, $O_2s^{22}p^4$, Ti $3p^{63}d^{34}s^1$ and Ta $5p^{66}s^{15}d^4$. The plane wave cut-off energy was500 eV and a 13 × 13 × 5 grid of Monkhorst-pack points wereemployed for geometry optimization of a 12-atom conventionalcell of anatase TiO_2. Geometry optimization was performed untilthe total energy difference reached 10-5 eV and the residualforces on atoms were less than 0.01 eV/Å. During the geometryoptimization, volume and shape of the cell as well as atomicpositions were relaxed.

To construct model structures for the anatase solid solution$(TiO_2)_{1-x}(TaON)_x$, we incorporated Ta and N atoms into lattice sites of a 12-atom anatase TiO_2, taking x value of 0.25, 0.5 and 0.75. Here, weconsidered all the possible lattice sites (see Fig. S1 and table S1, ESI) that Ta and N could occupy i.e., 4, 19 and 28 configurations forx = 0.25, 0.5 and 0.75, respectively, and found the model structure thatyields the lowest total energy through geometry optimizations for thevolume and shape of the cell as well as atomic positions. For anataseTaON, we studied the most plausible six atomic configurations withinthe 12-atom cell. The geometry relaxations were performed in thesame way as the solid solutions. The most stable configuration haveI41md symmetry, in agreement with the previous studies.[27,29,33,34]

3. Results and discussion

3.1 Optimized crystal structure

Fig. 1(a) shows the optimized anatase TiO_2. The lattice constants are a = 3.81Å, b = 3.81 Å, c = 9.69 Å, very similar with experimental results.[35] Figures 1(b) – 1(e) show the most stable configurations of the anatase $(TiO_2)_{1-x}(TaON)_x$ solid solutions at different concentrations, i.e., x = 0.25, 0.5, 0.75 and 1.0. Fig. 1(f) is the variation of the cell volume of the most stable $(TiO_2)_{1-x}(TaON)_x$ versus x value, indicating approximately linear cell increase when content of TaON increases in the solid solution because of the larger lattice constants of TaON than TiO_2, agreeing well with the experimental results.[29]

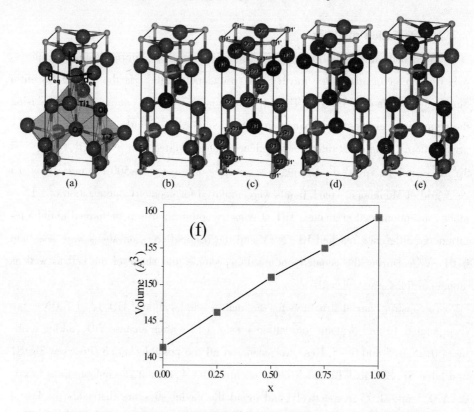

Fig. 1 Optimized crystal structure of $(TiO_2)_{1-x}(TaON)_x$ when x = 0(a), 0.25(b), 0.5(c), 0.75(d), 1.0(e), variation of the cell volume(f) versus TaON content. Red, purple, blue and brown balls represent oxygen, nitrogen, titanium and tantalum ions, respectively. The definition of the apical(dap) and equatorial(deq) bond lengths is indicated.

In the solid solution, metal T(T = Ti or Ta) coordinates with O or N to form octahedron. The cation – anion bond lengths are different along the apical (d_{ap}) and equatorial (d_{eq}) directions defined in Fig. 1. All the cation – anion bond lengths in the solid solution are listed in table 1. The enthalpy of mixing Ex for a cell with 12 atoms was obtained from the formula,

$$E_x = E_s - (1-x)E_{TiO_2} - x E_{TaON} \quad (1)$$

where E_s, E_{TiO_2}, and E_{TaON} are the calculated total energies of $(TiO_2)_{1-x}(TaON)_x$, TiO_2, and TaON, respectively, which are obtained from our DFT calculations for the 12 – atoms anatase models. The enthalpies of mixing of the solid solution are slightly positive ($E_{0.25}$ = 0.19eV, $E_{0.5}$ = 0.09eV and $E_{0.75}$ = 0.18 eV), indicating that the formation of solid solutions $(TiO_2)_{1-x}(TaON)_x$ is endothermic. Experimentally, this drawback can be overcome by precisely controlling epitaxial methods or soft chemical methods that are not bound by thermodynamic stability.

Table 1. Cation – anion bond lengths in different directions in the solid solution

x	Ti – O(Å)	Ti – N(Å)	Ta – O(Å)	Ta – N(Å)
0	1.95(d_{eq}) 2.01(d_{ap})			
0.25	1.93~1.99(d_{eq}) 1.96~2.13(d_{ap})	1.93(d_{ap})	1.97(d_{eq}) 2.08~2.09(d_{ap})	2.00(d_{eq})
0.5	1.96~2.01(d_{eq}) 2.22(d_{ap})	1.88(d_{ap})	1.97(d_{eq}) 2.04~2.09(d_{ap})	2.02(d_{eq})
0.75	1.97(d_{eq})	1.86(d_{ap})	1.97~1.99(d_{eq}) 2.22~2.30(d_{ap})	1.98~2.04(d_{eq}) 1.98~2.01(d_{ap})
1			2.01(d_{eq}) 2.15(d_{ap})	2.01(d_{eq}) 2.17(d_{ap})

3.2 Band alignment

In this section, we discuss the band alignment of the solid solutions $(TiO_2)_{1-x}(TaON)_x$ in comparison with their limiting phases TiO_2 and TaON. To estimate a band edge position of $(TiO_2)_{1-x}(TaON)_x$ with respect to that of TiO_2, we have performed a DFT calculation for an interface $(TiO_2)_{1-x}(TaON)_x/TiO_2$ following the method proposed by Janotti et al.[36-39] The interface model was built by stacking three unit cells of

$(TiO_2)_{1-x}(TaON)_x$ on top of three unit cells of TiO_2 along the c direction forming 72-atom supercell. Since the lattice constants of TiO_2 (p for short) and $(TiO_2)_{1-x}(TaON)_x$ (s for short) are different, we took average in-plane lattice constants, i.e. $a = (a_p + a_s)/2$ and $b = (b_p + b_s)/2$, for the construction of the interface models. Geometry relaxations were performed only for c axis with the averaged in-plane lattice constants fixed to take into account the Poisson ratio. The averaged electrostatic potentials over atoms ($V_{ref}[\cdots]$ interface) for $(TiO_2)_{1-x}(TaON)_x$ and TiO_2 in the interface model are estimated from the mid layer of each material which is far distant from the interface and can represent bulk region. We then define the difference of the averaged electrostatic potentials of the two materials as

$$DV_{ref} = V_{ref}[(TiO_2)_{1-x}(TaON)_x]_{interface} - V_{ref}[TiO_2]_{interface}.$$

Coming back to the bulk, the highest-occupied states $\varepsilon VBM[\cdots]$, averaged electrostatic potentials ($V_{ref}[\cdots]$ bulk), and band gaps $E_g[\cdots]$ for $(TiO_2)_{1-x}(TaON)_x$ and TiO_2 are also estimated from our DFT calculations using the 12-atom cell. Finally, the valence band and conduction band offsets of $(TiO_2)_{1-x}(TaON)_x$ with respect to TiO_2 are obtained from

$$De_{VBM} = (e_{VBM}[(TiO_2)_{1-x}(TaON)_x] - V_{ref}[(TiO_2)_{1-x}(TaON)_x]_{bulk})$$
$$- (e_{VBM}[TiO_2] - V_{ref}[TiO_2]_{bulk}) + DV_{ref},$$
$$De_{CBM} = De_{VBM} + E_g[(TiO_2)_{1-x}(TaON)_x] - E_g[TiO_2].$$

The obtained band alignment is shown in Fig. 2. It is understood that the VBM and CBM of $(TiO_2)_{1-x}(TaON)_x$ ($x = 0.25, 0.5,$ and 0.75) respectively shift upwards and downwards with respect to those of TiO_2, leading to band gap narrowing. In the case of a pristine anatase TaON, the CBM is much more negative than that of TiO_2, indicating a promising material for highly active photocatalytic water reduction. The obtained band alignment is shown in Fig. 2. It is understood that the VBM and CBM of $(TiO_2)_{1-x}(TaON)_x$ ($x = 0.25, 0.5,$ and 0.75) respectively shift upwards and downwards with respect to those of TiO_2, leading to band gap narrowing. In the case of a pristine anatase TaON, the CBM is much more negative than that of TiO_2, indicating a promising material for highly active photocatalytic water reduction.

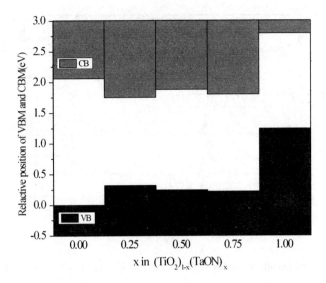

Fig. 2 The schematic band edges. The zero energy is at the valenceband maximum of pristine TiO_2.

3.3 Electronic structures

The relative band edge positions are rationalized by detailed analysisof electronic structures of the solid solutions. The density of states(DOS)for$(TiO_2)_{1-x}(TaON)_x$ are presented in Fig. 3 and Fig. S2(ESI), where the energy is referenced to the VBM of a pristine TiO_2 and theband offsets discussed in the previous paragraph are taken intoaccount for$(TiO_2)_{1-x}(TaON)_x$. In pristine TiO_2, the VBM and CBMmainly consist of O 2p and Ti 3d states,respectively. After one fourthof TiO_2 is substituted by TaON,the CBM is still dominated by Ti 3d orbitals,while a hybrid band of N 2p,O 2p,Ti 3d and Ta 5d is observedat the VBM. When the content of TaON increases to half in the solidsolution,the DOS is very similar with that of$(TiO_2)^{0.75}(TaON)^{0.25}$,except some Ta 5d orbitals appear at the CBM. The hybridization of N 2p,O 2p and cation d orbitals is pronounced near the VBM. Withincreasing TaON content in the solid solution,Ta 5d orbitals contribute further to the VBM and CBM character as shown in Fig. 3and Fig. S2(ESI).

The nature of chemical bonds among orbitals that is responsible forthe band edges can be visualized by partial electron density(squareof the wave function) associated with the energy states at the VBMand CBM as shown in Fig. 4. In TiO_2, the VBM is composed mainly of O 2p non – bonding states,while the CBM is dominated by the Ti 3dnon – bonding states as shown in Fig. 4(a),agreeing well with theresults reported by Asahi

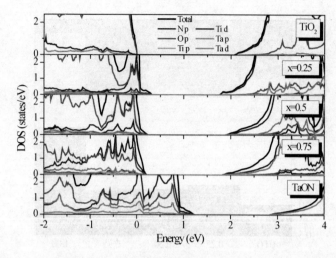

Fig. 3 Total density of states (TDOS) and partial density of states (PDOS) of $(TiO_2)_{1-x}(TaON)_x$ when $x = 0, 0.25, 0.5, 0.75$ and 1.0 near the VBM and CBM. The zero energy is at the valence band maximum of pristine TiO_2

and Y. Taga.[40] In $(TiO_2)0.75(TaON)0.25$, π bonding among Ti 3d states appears in Fig. 4(b), which is substantially different from the non-bonding states of Ti 3d in TiO_2 (Fig. 4(a)). Whereas, hybridization among N 2p, O 2p and Ti d_{xz}-states is observed at the VBM of $(TiO_2)_{0.75}(TaON)_{0.25}$. Similar band characters are found in the other concentrations ($x = 0.5$ and 0.75) as shown in Fig. 4(c) and 4(d). Cation-anion or cation-cation π bond formation is promoted as increase in the TaON content. The electron clouds at the CBM are connected forming a zigzag-like character along d-d π bonds that distribute all over the crystal, while those at the VBM are connected in a xy plane.

According to Jeremy K. Burdett et al.,[35] TiO_6 octahedron distortion originated from some inter-constraint factors, such as Coulomb repulsion between neighboring Ti atoms, repulsion between two oxygen atoms on the edges shared by two octahedron (Fig. 1(a)) and attraction between Ti and O. Since ionic radii of four-coordinated N^{3-} is larger than four-coordinated O^{2-} (1.46Å vs. 1.38Å), it is anticipated that three-coordinated N3- is larger than three-coordinated O^{2-}. Meanwhile, six-coordinated Ta^{5+} is larger than six-coordinated Ti^{4+} (0.64Å vs. 0.605Å).[41] As a result, the cell volume increases with increasing the TaON ratio in the solid solution. Furthermore, the Coulomb repulsion between N and O exceeds that between two O atoms, causing the longer N-O distance (2.53Å) than O-O distance (2.48 Å). This repulsion causes the octahedron containing this O atom(denoted as O* in Figures 4(b) –4(d))to have a lower distortion alonga Ti – O* – Ti network, namely, its angle closer to 180 degree than inTiO_2. With increasing the TaON ratio in the solid solution, thedistortion continues to decrease. The elongation of anion – aniondistance in a quadrangle containing two cations and two anions, alsocauses the cations approach to each other along the z direction, resulting in the attraction of dxyorbitals forming the zigzag – like d – d π bonding network. The π bonding significantly lowers the CBM in thesolid solutions compared to the case in TiO_2 in which the CBM consistsof non – bonding Ti d states. The rise of the VBM in the solid solutionsmainly originates from the fact that the atomic eigenvalues of 2p states are higher in nitrogen than in oxygen. This is still the case inour solid solutions even after the hybridization with d orbitals from Tior/and Ta. The extended 2p states around N in the solid solutionspromote the formation of π bonding with cation d states, whichsomewhat stabilizes the system shifting the VBM downwards, although the resulted VBM is still higher than that in a pristine TiO_2. As a result of the combined two effects of the rise of VBM and declineof CBM, the solid solutions have narrower band gaps than that of TiO_2.

In Fig. 5(a) –5(e), band structures along the symmetry directions forthe Tetragonal Lattice Brillouin zone(Fig. S3, ESI) for TiO_2, $(TiO_2)_{1-x}(TaON)_x$ (x = 0.25, 0.5, and 0.75), and TaON are shown. The pristineTiO_2 has an indirect band gap of about 2.06eV for a transition from M to Γ as exhibited in Fig. 5(a), much smaller than an experimentalvalue 3.2eV,[42] due to the well – known shortcoming of GGA.[43] Asmentioned earlier, in$(TiO_2)0.75(TaON)0.25$, the VBM and CBM shiftupwards and downwards, respectively, leading to a narrower bandgap(1.43eV) compared to the pristine TiO_2(Fig. 4(b)), while theindirect nature is barely remained for a transition from Z to Γ.

Further increase in the TaON content in the solid solution turns theband transition into direct at Γ. This indicates that the solid solutionwith a higher TaON content is more suitable for light absorption. Notably, the incorporation of TaON into TiO_2 lattice does not induceany localized energy levels in a forbidden band, unlike lowconcentration N – doped TiO_2. R. Long et al. studied the(N, Ta) – codoped TiO_2 via DFT.[14] Their results indicate that codoping N and Tacan effectively enhance N concentration and besides, lead to theformation of continuum – like fully occupied N 2p – Ta 5d hybridizedstates a-

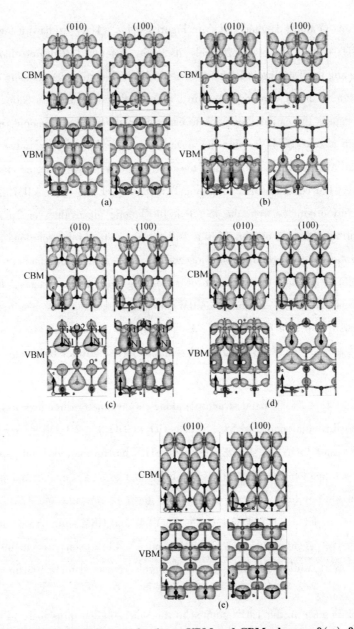

Fig. 4 The partial electron density at VBM and CBM when x = 0(a), 0.25 (b), 0.5(c), 0.75(d), 1.0(e) at a 0.01 electrons/Å^3 isosurface level. The left and right panel indicates the (010) and (100) plane, respectively.

bove the top of the valence band as well as Ta 5d orbitalslocated at the bottom of the conduction band. However, such animpurity state often produces a small quantity of occupation, resulting in the declined photocatalytic activity under UV irritationalthough it

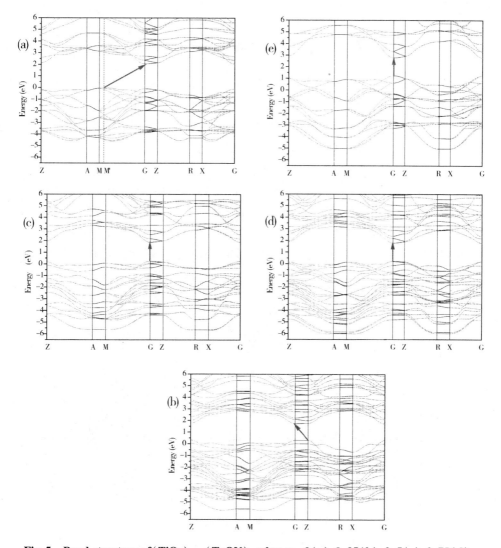

Fig. 5 Band structure of $(TiO_2)_{1-x}(TaON)_x$ when $x = 0(a), 0.25(b), 0.5(c), 0.75(d)$, and $1.0(e)$, plotted along the symmetry directions for the Tetragonal Lattice Brillouin zone (Fig. S3, ESI). Red arrow indicates the VBM and CBM. The zero energy is at the valence band maximum of pristine TiO_2.

benefits the activity under visible light.[44,45] The localized energy levels in the forbidden band usually tend to be a recombination center between photogenerated carriers, and the absorption coefficient depends on the concentration of dopants. In contrast, in $(TiO_2)_{1-x}$ $(TaON)_x$ solid solutions, it is N 2p and O 2p orbitals To more thoroughly understand the molecular orbitals of the solid solution, we analyze the PDOS, taking $(TiO_2)0.5(Ta$-

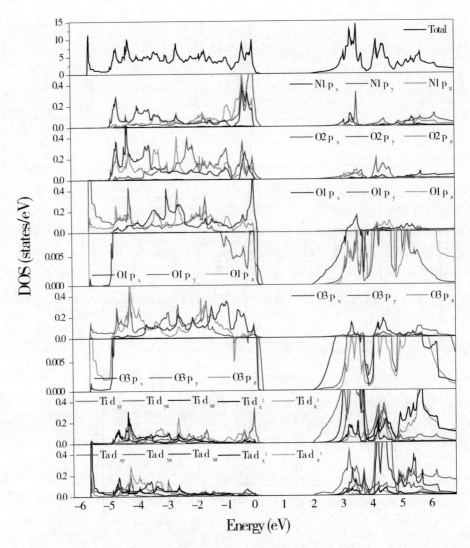

Fig. 6　PDOS of (TiO$_2$)0.5(TaON)0.5.

ON)0.5 as an example. As can be seen from the crystal structure shown in Fig. 1(c), O1 and O1' occupy the identical sites. So do O2 and O2', O3 and O3', two nitrogen atoms, two titanium atoms and two tantalum atoms. Hence, we choose either one to analyze the PDOS. O2 and N1 atoms form an edge shared by two octahedrons and reside in a same quadrangle, having a similar profile of PDOS, i.e. the highest occupied state consists of anion py states as shown in Fig. 6. In anatase structure, an anion is coordinated with three cations in a planner configuration, which is the characteristic of O p – Ti d hybrid states, to form σ bonds and either px, py, or pz state, which is orientally perpendicular to the plane, forms a π bond with cation d states. In the case of O2 and N1, py state is hybridized with Ti dyz through π interactions. The N1 py – Ti dyz π bonds are clearly observed in the partial electron density (Fig. 4(c)). This is the reason for the appearance of py at the VBM and its resonance with Ti dyz in PDOS. The intensity of DOS at the VBM is slightly higher in N1 than in O2 because N 2p states are energetically higher than O 2p states. Similarly, O1 and O3 reside at a quadrangle in yz plane having px states dominated near the top of the valence band through the hybridization with Ti dxz, although a small amplitude of py appears right at the top of the valence band following the upward shift of Ti dyz that forms σ bonds with py. The difference between O1 and O3 in the DOS profile originates from cation coordination, i. e. O1 coordinates with two Ta atoms, while O3 coordinates with two Ti atoms. Ta 5d states are energetically higher than Ti 3d states and their interactions with oxygen p states are limited. In O1, therefore, O p states are intact against hybridization with Ta d states remaining at the VBM, whereas in O3, the hybridization between O p and Ti d is significant, broadening the oxygen band. In fact, a strong σ bond of O3 py with Ti dyz is observed in Fig. 4(c). At the CBM, Ti dxy and Ta dxy are also resonant due to their π bonding as discussed earlier, which results in the formation of the lower energy states compared to nonbonding Ti dxy states in a pristine TiO_2.

3.4 Some configurations possessing very narrow band gap

In the above two sections, we presented the crystal structure and electronic structure of most stable configurations in the $(TiO_2)_{1-x}(TaON)_x$ solid solutions. In addition, we found that some models exhibited very small band gap.

For $(TiO_2)0.5(TaON)0.5$, if the atoms are arranged as Fig. 7(a), the band gap is only 0.4 eV as shown in Fig. 7(e). DOS and the partial electron densities of the VBM and CBM in Fig. 7 indicate that the CBM is mainly composed of Ti 3d orbits while the VBM majorly consists of N 2p orbitals, substantially different from the cases shown in Fig. S2(c) (ESI) and Fig. 4(c) in which O 2p orbits contribute approximately half to the VBM besides minor contribution from d states of titanium or tantalum. For $(TiO_2)_{0.5}(TaON)_{0.5}$ with total 19 models, some models have even smaller band gaps (table S1(b), ESI). Similarly with the case in Fig. 7, the VBM majorly consists of N 2p orbits, and CBM is composed by T dxy orbitals. Several models of $(TiO_2)0.25(TaON)0.75$, also possess small band gaps (table S1(c), ESI), have the similar characteristic of the DOS.

After thoroughly analyzing the configurations that have much smaller band gaps

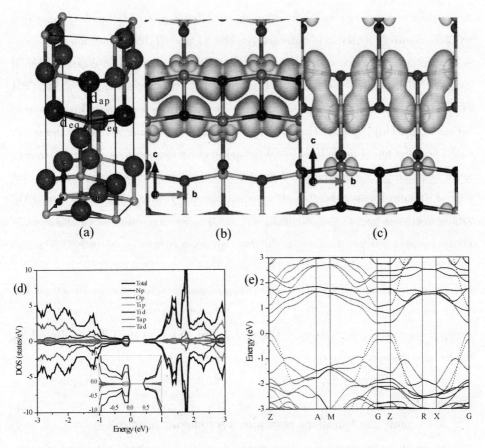

Fig. 7 One model of $(TiO_2)_{0.5}(TaON)_{0.5}$ that has a small band gap: thecrystal structure (a), partial electron density in the VBM(b) and CBM(c) at a 0.01 electrons/Å^3 isosurface level, DOS(d) and band structure(e). The energy band edge in this section was not assigned using theabove interface model because we majorly concerned band gapvalue.

than the others, we found one common featuresfor these configurations: two N atoms form edge shared by twooctahedrons. This is different from the most stable configurationshown in Fig. 1, where N and O form edge shared by two octahedrons. As mentioned above, N has broader electron cloud distribution than O, causing larger Coulomb repulsion. The balance between therepulsions from N – N and the nearby cations leads to the declineddistortion in these two octahedrons along the equatorial direction, and hence the N – T – N – T quadrangle approaches to rectangle. As aresult, N 2p and Ta 6p form σ bond along the equatorial direction, asshown in Fig. 7(b) and Fig. 8. Because energy level of Ta 6p is higherthan that of Ta 5d, the Ta 6p – N 2p bonding states form a σ bandabove the π band originating from Ta 5d – N 2p bonds, which substantially raises the VBM. On the other hand, similar with the configuration in Fig. 4, the CBM is composed by d – d bonding states with a lower energy than the non – bonding states. These two aspects contribute together to the small band gap of the solid solution (Fig. 7(e)). We calculated the band gaps and the total energy difference of all the configurations in comparison with the most stable ones as shown in table S1 (ESI). The most stable configurations have the lowest total energy and the largest band gap. The structures, which possess exceptionally narrow band gap, have the highest energy in most cases. This small band gap impedes achieving high photocatalytic activity, but fortunately, they have high enthalpy of mixing indicating that they have little chance to appear during the experiments. Maybe, these configurations can be obtained, for the special purpose, by employing advanced experimental methods such as the soft chemical synthesis.

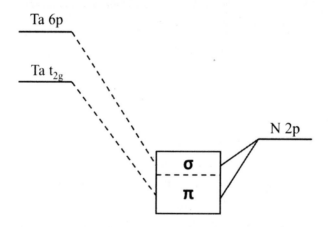

Fig. 8 Schematic molecular orbitals near the VBM in the configurations where two N atoms form an edge shared by two octahedrons

4. Conclusions

TaON and anatase TiO_2 can form stable solid solution, while the cell volume increases approximatedly linearly and the inter – band transitions turn from in – direct to direct transitions with increasing TaON content in the solid solution. Introducing TaON into TiO_2 causes the shortened distance between some cations and leads to d – d metallic bonding states instead of non – bonding states, lowering the CBM. Hybridization of N 2p, O 2p and d orbitals from Ti or/and Ta upraises the VMB without forming any localized energy band in the forbidden band. The solid solutions possess higher charge – carrier mobility than the prestine TiO_2 because of the reduced hole effective mass. Anatase TaON has much negative CBM, indicating its potential as a visible – light phtocatalyst to reduce water. We have found some peculiar configurations with small band gap, in which two N atoms form an edge shared by two octahedrons. The higher Coulomb repulsion of N – N in comparison with O – O and O – N almost deletes the distortion in the T – N3 (T = Ti or Ta) plane, causing high – energy σ bond of N p – T p at VBM. In summary, we propose to obtain visible – light anatase photocatalyst by forming solid solution of TaON and TiO_2, realizing high – concentration doping and avoiding the localized energy state.

Acknowledgements

We would like to thank Y. Aoki and J. Ye for useful discussion. This work was supported by Ph. D Programs Foundation of the Ministry of Education of China (Grant No. 20121102110027), and the National Science Foundation of China (Grant Nos. 91222110 and 51472013).

References

[1] A. Fujishima and K. Honda, Nature 1972, **238**, 37.

[2] R. Asahi, T. Morikawa, T. Ohwaki, K. Aoki and Y. Taga, Science 2001, **293**, 269.

[3] S. U. M. Khan, M. Al – Shahry, and W. B. I. Jr, Science 2002, **297**, 2243.

[4] T. Umebayashi, T. Yamaki, H. Itoh, and K. Asai, Appl. Phys. Lett. 2002, **81** (3), 454.

[5] J. C. Yu, J. G. Yu, W. K. Ho, Z. T. Jiang and L. Z. Zhang, Chem. Mater. 2002, **14**, 3808.

[6] S. In, A. Orlov, R. Berg, F. García, S. Pedrosa – Jimenez, M. S. Tikhov, D. S. Wright and R. M. Lambert, J. Am. Chem. Soc. 2007, **129**, 13790.

[7] C. D. Valentin, G. Pacchioni and A. Selloni, Phys. Rev. B 2004, **70**, 085116.

[8] Z. S. Lin, A. Orlov, R. M. Lambert and M. C. Payne, J. Phys. Chem. B 2005, **109**, 20948.

[9] J. B. Varley, A. Janotti and C. G. Van de Walle, Adv. Mater. 2011, **23**, 2343.

[10] W. Choi, A. Termin and M. R. Hoffmann, J. Phys. Chem. 1994, **98**, 13669.

[11] J. Choi, H. Park and M. R. Hoffmann, J. Phys. Chem. C 2010, **14**, 783.

[12] M. E. Kurtoglu, T. Longenbach, K. Sohlberg and Y. Gogotsi, J. Phys. Chem. C 2011, **115** (35), 17392.

[13] M. Y. Xing, Y. M. Wu and J. L. Zhang, Chen F., Nanoscale 2010, **2**, 1233.

[14] R. Long and N. J. English, Chem. Phys. Lett., 2009, **478**, 175.

[15] T. M. Breault and B. M. Bartlett, J. Phys. Chem. C 2012, **116**, 5986.

[16] S. Sato, Chem. Phys. Lett. 1986, **123**, 126.

[17] T. Okato, T. Sakano and M. Obara, Phys. Rev. B 2005, **72**, 115124.

[18] T. L. Chen, Y. Hirose, T. Hitosugi and T. Hasegawa, J. Phys. D: Appl. Phys. 2008, **41**, 062005.

[19] K. Maeda, and K. Domen, J. Phys. Chem. C 2007, **111**(22), 7851.

[20] B. F. Gao, T. M. Lim, D. P. Subagio and T. Lim, Applied Catalysis A: General 2010, **375**, 107.

[21] R. Schiller, C. K. Weiss and K. Landfester, Nanotechnology 2010, **21**, 405603.

[22] K. Maeda, T. Takata, M. Hara, N. Saito, Y. Inoue, H. Kobayashi and K. Domen, J. Am. Chem. Soc. 2005, **127**, 8286.

[23] K. Maeda, K. Teramura, D. L. Lu, T. Takata, N. Saito, K. Inoue and K. Domen, Nature 2006, **440**, 295.

[24] D. F. Wang, T. Kako and J. H. Ye, J. Am. Chem. Soc. 2008, **130**, 2724.

[25] S. X. Ouyang and J. H. Ye, J. Am. Chem. Soc. 2011, **133**, 7757.

[26] I. Tsuji, H. Kato, H. Kobayashi and A. Kudo, J. Am. Chem. Soc. 2004, **126**, 13406.

[27] T. Lüdtke, A. Schmidt, C. Göbel, A. Fischer, N. Becker, C. Reimann, T. Bredow, R. Dronskowski, and M. Lerch, Inorg. Chem., 2014, **53**, 11691.

[28] A. Suzuki, Y Hirose., D. Oka, S. Nakao, T. Fukumura, S. Ishii, K. Sasa, H. Matsuzaki and T. Hasegawa, Chem. Mater. 2014, **26**, 976.

[29] J. Grins, J. Europ. Ceram. Soc. 1997, **17**, 1819.

[30] G. Kresse and D. Joubert, Phys. Rev. B 1999, **59**, 1758.

[31] G. Kresse and J. Furthmüller, Phys. Rev. B 1996, **54**, 11169.

[32] J. P. Perdew, K. Burke and M. Ernzerhof, Phys. Rev. Lett. 1996, **77**, 3865.

[33] T. Bredow, M. W. Lumey, R. Dronskowski, H. Schilling, J. Pickardt, and M. Z. Lerch, Anorg. Allg. Chem. 2006, **632**, 1157.

[34] H. Wolff, M. Lerch, H. Schilling, C. Bähtz and R. Dronskowski, J. Sol. Stat. Chem., 2008, **181**, 2684.

[35] J. K. Burdett, T. Hughbanks, G. J. Miller, J. W. Richardson, Jr. and J. V. Smith, J. Am. Chem. Soc., 1987, **109**, 3639.

[36] C. G. Van de Walle and R. M. Martin, J. Vac. Sci. Technol. B 1987, **5**, 1225.

[37] C. G. Van de Walle and R. M. Martin, Phys. Rev. Lett. 1989 **62**, 2028.

[38] A. Janotti and C. G. Van de Walle, Phys. Rev. B 2007, **75**, 121201.

[39] N. Umezawa, A. Janotti, P. Rinke, T. Chikyow and C. G. Van deWalle, Appl. Phys. Lett., 2008, **92**, 041104.

[40] R. Asahi and Y. Taga, Phys. Rev. B 2000, **61**, 7459.

[41] R. D. Shannon. Crystal Physics, Diffraction, ActaCrystallogr. , Sect. A 1976, **32**, 751.

[42] H Tang. , H. Berger, P. E. Schmid, F. Levy and G. Burri, Solid StateCommun. , 1993, **87**, 847.

[43] R. O. Jones and O. Gunnarsson, Rev. Mod. Phys. 1989, **61**, 689.

[44] J. Wang, D. N. Tafen, J. P. Lewis, Z. L. Hong, A. Manivannan, M. J. Zhi, M. Li and N. Q. Wu, J. Am. Chem. Soc. 2009, **131**, 12290.

[45] D. N. Tafen, J. Wang, N. Q. Wu and J. P. Lewis, Appl. Phys. Lett. 2009, **94**, 093101.

注:本文曾发表在 2015 年《Physical Chemistry Chemical physics》

基于 NSGA – II 算法的管道清灰
机器人变径机构优化

罗海玉　张淑珍*

管道清灰机器人变径机构尺度影响机构的运动性能及驱动性能,变径机构尺度优化可有效解决尺寸综合问题。在变径机构运动学和力学分析基础上,提出变径机构多目标尺度综合,以变径机构关键零件受力和驱动件运动范围为优化目标建立优化模型,基于快速含有精英策略的非支配排序遗传算法(Non – dominated sorting genetic algorithm II, NSGA – II)求解多目标优化 Pareto 最优解,数值计算结果表明多目标优化后的变径机构在力学性能和运动范围上优于经验设计,不需重复计算可根据设计要求和工程经验权衡选取满足不同要求的优化结果。

1　引言

管道在使用过程中,由于各种外界因素的影响,会形成各种各样的管道故障与管道损伤。然而,管道所处的环境是人们不易直接达到或不允许人们直接进入的,检修及清洗难度很大,因此最有效的方法之一就是利用管道机器人来实现管道内的在线检测、维修和清洗。

目前的管道机器人,如 Heli – Pipe 管道机器人[1]和法国 Cedric Anthierens 等为检测核电站蒸汽发生器管道而研制的"电动 – 气动"驱动的蠕动式管道机器人[2]等,都是为特定直径的管道专门设计的。这种管道机器人不能在管径一定范围内连续变化的管道内实施作业。哈尔滨工业大学的邓宗全等研制的六轮独立驱动管道机器人[3],采用弹簧封闭力机构,实现适应管径变化的功能,且每个行走轮都配有独立驱动电机和传动机构。韩国 Choi H R 等开发出的用于天然气管道检测的管道机器人[4],当工作管道管径变化范围较大时,管道机器人的移动机构

---

* 作者简介:罗海玉(1970—　),男,甘肃天水人,天水师范学院教授、硕士,主要从事机构设计及智能 CAD 的研究。

的牵引力将不再稳定,机器人的工作可靠性将受到很大的影响。此外,变径机构与传动机构相互独立,使得机器人机构相对复杂,尺寸较大。上海大学陆麒等研制的管道机器人管道公称直径为100 mm,适应管径变化范围为96 mm~102 mm的管道。

金属冶炼厂的烟气输送管道,时间一长在管道底部会形成厚厚的烟灰堆积层,为不影响制酸质量及金属物料的损失,需定期进行清理。传统的人工清理方式,由于管道直径过小,人无法直立,毒性及辐射物质过多,工作环境危险性极大,因此采用自动化清灰方式,研制适合 Φ700mm~Φ1000mm 管径的管道清灰机器人[5]。

管道清灰机器人三维结构模型如图1所示,机器人由移动机构、变径机构、机械手、吸尘器、卸料机构、信号采集设备、机器人电控装置、收放线缆装置等组成。移动机构由3组均布履带足组成,履带足内置电机并独立驱动可实现移动机构速度协调控制。卸料机构采用曲柄连杆机构,利用机构死点位置使卸料门闭合,曲柄转动,卸料门打开。机械手为3自由度平行四边形操作臂,手臂末端保持与管壁平行,手臂末端安装吸尘器头,通过吸尘软管将灰尘收集到集尘箱中。变径机构不但起到变径目的,同时还提供履带足在管壁行走时的附着力以及变径的驱动力,变径机构设计的优劣影响行走机构性能及在管道中行走的灵活性,因此有必要对管道机器人变径机构进行尺度优化综合。

1—履带足 2—变径机构 3—箱体 4—机械手
5—吸尘器头 6—卸料机构 7—丝杆螺母

图1 管道清灰机器人模型

2 变径机构

变径机构如图 2 所示采用曲柄滑块机构,滑块移动使曲柄转动推动履带足升高或降低达到变径目的。杆件 2 为一柔性支撑杆组,当行走机构受到管道空间的约束,径向发生微小变化时,在主动径向调节来不及响应时,连杆系统可以带动推杆压缩弹簧,来抵消管径变化,使行走机构在径向具有一定的柔性。

管道清灰机器人作业环境管径为 Φ700mm～Φ1000mm 的管道,因此设计的机器人变径机构要有适应 0～300mm 之间的变径能力。变径机构安装在 3 个沿圆周均布的履带足上,每组变径机构至少要满足 0～150mm 的变径范围,变径范围是由曲柄 2 的转角 θ_2 决定的,只要支撑履带足的杆件 $l_5 \geq 150$ mm 即可满足变径范围,但在满足变径范围的前提下,还需考虑变径机构的驱动性能和运动范围等其他性能是否达到最优。

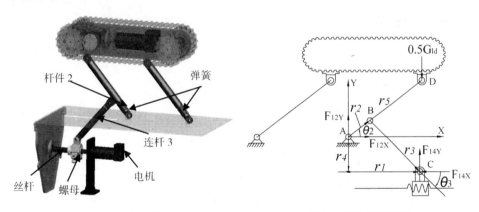

图 2　变径机构　　　　　　　图 3　变径机构简图

三个履带足变径机构相同,机构简图及受力分析如图 3 所示,写出撑开机构闭环矢量方程

$$R_4 + R_1 = R_2 + R_3 \tag{1}$$

将矢量方程分解到 X 和 Y 坐标轴上,有

$$r_1 = r_2\cos\theta_2 + r_3\cos\theta_3$$
$$r_4 = r_2\sin\theta_2 + r_3\sin\theta_3 \tag{2}$$

力平衡方程

$$\begin{cases} F_{12X} - F_{14X} = 0 \\ F_{12Y} - 0.5G_{ld} + F_{14Y} = 0 \\ F_{14Y}(r_3\cos\theta_3 + r_2\cos\theta_2) - \\ 0.5G_{ld}r_5\cos(\theta_2 + \varphi) - F_{14X}r_4 = 0 \end{cases} \tag{3}$$

其中 G_{ld} 表示履带足的重量。

由式(2)和式(3)可得出

$$F_{14x} = \frac{0.5 G_{ld} r_5 cos \theta_2}{r_2 cos \theta_2 \left(\frac{r_4 - r_1 sin \theta_2}{\sqrt{r_3^2 - (r_4 - r_2 sin \theta_2)^2}} \right) - r_2 sin \theta_2} \tag{4}$$

3 变径机构优化模型

变径机构优化目的是确定机构尺寸使变径机构在撑开履带足时施加在滑块上的力最小,即 F_{14X} 最小,同时在满足变径范围的条件下,滑块的行程最短,即 r_1 最小。考虑管道机器人车身及行走机构结构限制,设计滑块偏置距离为 $r_4 = 140mm$,履带足重 $G_{ld} = 12.5$ kg。设计变量为

$$X = [r_2, r_3, r_5]^T = [x_1, x_2, x_3]^T$$

3.1 目标函数

目标函数为滑块水平受到的分力 F_{14X} 和滑块行程范围 r_1,F_{14X} 随曲柄转角 θ_2 变化,因此为 $max(F_{14X})$。

$$f_1(X) = max \left(\frac{0.5 G_{ld} x_3 cos \theta_2}{x_1 cos \theta_2 \left(\frac{r_4 - x_1 sin \theta_2}{\sqrt{x_2^2 - (r_4 - x_1 sin \theta_2)^2}} \right) - x_1 sin \theta_2} \right) \tag{5}$$

$$f_2(X) = \left(\sqrt{r_3^2 - r_4^2} + r_2 \right) - \left(\sqrt{r_3^2 - (r_4 + r_2)^2} \right) \tag{6}$$

考虑变径机构结构约束取 $0° \leq \theta_2 \leq 90°$。

3.2 约束条件

考虑变径机构结构和变径范围要求,变量范围有

$$\begin{cases} 30 \leq r_2 \leq 200 \\ (r_2 \leq r_5 \leq 300) \cup (160 \leq r_5) \\ 150 \leq r_3 \leq 320 \end{cases} \tag{7}$$

考虑曲柄存在条件有: $\quad r_2 + r_4 \leq r_3 \tag{8}$

写出约束条件

$$\begin{cases} x_1 - x_2 \leq -140 \\ x_1 - x_3 < 0 \end{cases} \tag{9}$$

优化问题为多目标约束规划问题,对多目标优化问题,由于存在目标之间的无法比较和冲突现象,不一定存在所有目标上都是最优的解,一个解可能在某个目标上是最好的,但在其他目标上是最差的。因此有多个目标时,通常存在一系

列无法简单比较的解,称作非支配解或 Pateto 解。传统的多目标求解方法是将多目标转化为单目标,往往只能得到一个解,而遗传算法的内在并行机制及其全局优化的特点适合于多目标优化问题的解决。

4 变径机构多目标优化

4.1 NSGA–II 算法

NSGA–II 算法是一种带精英策略的非支配排序遗传算法[6-8],该算法既有良好的分布性又有较快的收敛速度,得到广泛应用。NSGA II 的精英保留策略使用($\lambda + \mu$)选择,包含了最好的父代和子代个体,这种机制使新一代种群比前一代种群更有效,效果更好,成为许多其他多目标进化算法(MOEA)的比较对象。NSGA–II 算法流程如图 4 所示。

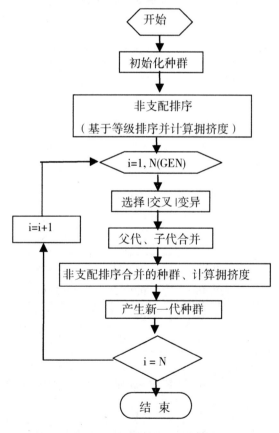

图 4 NSGA–II 算法流程

(1)随机产生种群规模为 N 的初始种群 P_0,用快速非支配排序法对初始种群

进行排序,对同一前沿面解集计算每个解的拥挤度并按拥挤度排序;

(2)在初始种群中选择数量为 pool-size 的个体作为父代种群 $Parent_0$,并通过交叉、变异产生子代种群 C_0;

(3)将初始种群 P_0 与子代种群 C_0 合并组成规模为(N + pool - size)大小的中间种群,对中间种群进行快速非支配排序,根据排序结果选取 N 个个体作为新一代种群 P_1;

(4)在种群 P_1 中通过遗传算子产生子代种群 C_1;

(5)重复步骤(3)~(4),直到算法设置的最大迭代代数。

4.2 求解过程及优化结果

取种群大小 200,运行代数 800 代,模拟二进制交叉分布指数为 $\eta_c = 20$,多项式变异分布指数为 $\eta_m = 20$。NSGA-II 算法计算的 Pareto 解如图 5 所示,将 Pareto 解集按目标 $f(x_1)$ 排序,即滑块水平受力最小排序,限于篇幅,从中选取 16 组解列在表 1 中。

表 1 Pareto 最优解中选取 16 组解

序号	r_2 mm	r_3 mm	r_5 mm	$f_1(x)$	$f_2(x)$
1	100.51	319.99	160.04	1.076	177.17
2	99.73	319.99	160.04	1.084	175.52
3	98.65	319.99	160.04	1.096	173.23
4	97.68	320.00	160.04	1.107	171.17
5	96.90	319.99	160.05	1.116	169.54
6	96.01	320.00	160.00	1.126	167.66
7	95.99	320.00	160.00	1.126	167.62
8	95.18	320.00	160.00	1.136	165.92
9	94.06	319.46	160.02	1.15	163.78
10	93.08	319.54	160.00	1.162	161.73
11	92.28	319.47	160.02	1.172	160.10
12	91.44	319.47	160.01	1.183	158.38
13	89.87	320.00	160.19	1.205	155.00
14	87.89	320.00	160.19	1.232	150.99
15	84.72	319.97	160.09	1.277	144.67
16	84.37	320.00	160.09	1.283	143.95

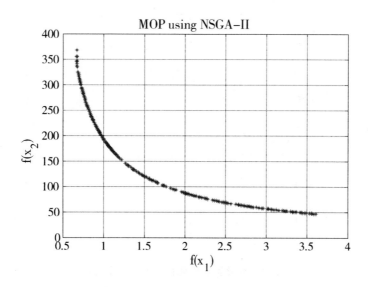

图 5　变径机构多目标 Pareto 解

从图 5 和表 1 可以看出变径机构两个优化目标下的优化解集,当目标 $f_1(x)$ 最优时即滑块受力最小时,滑块运动范围并未取得最优即滑块行程最小,如果滑块行程加长,使得机箱结构加长,管道机器人在转弯时影响机器人运动灵活性;当目标 $f_2(x)$ 最优,即滑块运动范围最小时,目标 $f_1(x)$ 未取得最优。

从表 1 还可看出优化变量中 r_3 取得变量范围的上限值,r_5 取得变量范围的下限值,根据结构设计要求和工程经验选取表 1 中的第 5 组解并取整得 $r_2 = 97$ mm,$r_3 = 320$ mm,$r_5 = 160$ mm。用优化后的杆件尺度和优化前由经验所得的尺寸 $r_2 = 92$ mm,$r_3 = 238$ mm,$r_5 = 272$ mm 对优化目标进行计算,图 6 为优化前后目标 1 结果。

图 6 中带"∗"标记的曲线为优化前的目标 1 结果,带"+"标志为优化后的目标 1,可以看出优化后的结果明显优于优化前目标。计算优化目标 2 得到优化前为 231 mm,优化后为 169 mm,优化结果也明显优于优化前。

4　结论

管道清灰机器人变径机构多目标优化过程可以看出优化目标对优化结果的影响。①优化目标为单目标时,取得的优化结果不能满足其他的优化目标;②优化目标为多目标时,借助 NSGA-II 算法得到多目标的全部 Pareto 最优解,不需重

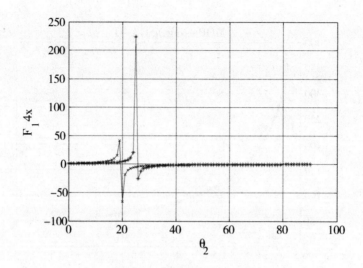

图6　优化前后优化目标1的计算结果

复计算可根据设计要求和工程经验权衡选取满足不同要求的优化结果,得到管道清灰机器人变径机构最优的尺度和良好的运动性能。

参考文献

[1] Horodinca M, Doroftei I, Mignon E, et al. A simple architecture for in‑pipe inspection robots[C]. Colloquium on Mobile and Autonomous Systems, 10 Years of the Fraunhofer IFF, Magdeburd, Germany, 2002:61‑64.

[2] Anthierens C, Ciftci A, Betemps A. Design of an electro pneumatic micro robot for in‑pipe inspection[C]. Industrial Electronics, 1999. ISIE '99. Proceedings of the IEEE International Symposium on Bled, Slovenia, 1999:968‑972.

[3] 邓宗全,陈军,姜元生,等. 六独立轮驱动管内检测牵引机器人[J]. 机械工程学报, 2005, 41(9):68‑72.

[4] Chio H R, Ryew S M, Robotic system with active steering capability for internal inspection of urban gas pipelines[J]. Mechatronics, 2002, 12(5):716‑736.

[5] 张淑珍,黄建龙,李春玲,改进型管道清灰机器人系统研究[J],机械与电子,2009. 12, No. 12, 62‑64.

[6] Sergiu‑Dan Stan,, Vistrian Mătieú, Radu Bălan, Genetic Algorithms Multi‑objective Optimization of a 2 DOF Micro Parallel Robot, Proceedings of the 2007 IEEE International Symposium on Computational Intelligence in Robotics and Automation Jacksonville, Hsinchu, June 20‑23, 2007:522‑527.

[7] DEB K,PRATAP A,AGARWAL S,et.a1. A fast and elitist multi-objective genetic algorithm:NSGA-II [J]. IEEE Transactions on Evolutionary Computation,2002,6(2):182-197.

[8]张超勇,董星,王晓娟,李新宇,刘琼,基于改进非支配排序遗传算法的多目标柔性作业车间调度,机械工程学报[J],Vol.46,No.11:156-166,2010,6.

注:本文曾发表在2013年《机械科学与技术》第10期上

一种新型并联太阳能跟踪机构研究

罗海玉[*]

设计了一种新型的并联太阳能跟踪机构,分析了并联跟踪机构的运动特性,给出位置逆解。依据太阳方位分析了跟踪性能,给出了并联平台转角与太阳位置的关系,并分析了并联机构支链受力和驱动性能。该机构结构简单,刚度大,运动范围大,可有效承载大面积光伏板并实现高效跟踪。

太阳能作为一种清洁的可再生能源,受到广泛关注。其中光伏发电就是利用光生伏特效应原理,利用太阳电池将太阳光能直接转化为电能。而提高太阳能利用率降低发电成本,有效的办法一是提高电池的能量转换率,二是提高太阳能的接收效率,前者属于能量转换领域,后者属于太阳跟踪技术。自动跟踪是根据太阳的实际位置,光伏平台在南北垂直轴和东西水平轴上分别随太阳的方位角和高度角的变化,利用传动机构完成相应的跟踪转动。理论研究表明[1]:太阳能自动跟踪系统比同等条件下的非跟踪系统能量接受效率要高大约37.7%,因此研究自动跟踪技术对于提高太阳能利用率的意义很大。

目前的跟踪机构可分为单轴跟踪和双轴跟踪[2-4]。单轴太阳能跟踪装置结构简单,制作费用低,太阳能接收效率比固定式装置有较大提高。单轴跟踪装置一般只能实现东西方向的太阳跟踪,而不能对太阳南北方向的倾角变化进行自动跟踪,太阳能接收率相对来说仍较低。现有的双轴跟踪机构种类很多,有传统的二轴跟踪机构、极轴跟踪机构、三自由度并联跟踪机构等。传统的二轴跟踪机构、极轴跟踪机构属于串联式机构跟踪,即一个角度方向的驱动系统的重量必须由另一角度方向的末端驱动装置来承担,串联式增加了末端驱动装置的功率,同时体积庞大,浪费材料和能源。并联机构具有刚度大,结构稳定,承载力强,精度高,易

---

[*] 作者简介:罗海玉(1970—),男,甘肃天水人,天水师范学院教授,硕士,主要从事机构设计及智能CAD的研究。

于实现实时运动控制等优点,在医疗器械、机床、太阳跟踪等领域多有应用[5-6]。应用在太阳自动跟踪上的并联机构有张顺心教授提出的三自由度并联球面跟踪机构[7]和贺新升等[8]提出的一种三自由度并联机构。张提出的并联球面跟踪机构具有结构紧凑、刚度较高、工作空间较大等特点,较好地解决了传统二轴跟踪机构跟踪范围小,精度较低的问题,实现了对太阳的全方位跟踪,大大提高了接收效率。但由于整个装置的重力都落在3个电机上,所以电机的自身耗电量大,输出有效发电量小。贺新升等提出并联跟踪机构,解决了机构自身的电机耗电量较大的问题,输出的有效电量较多,但机构复杂。本文提出一种新型并联太阳能跟踪机构,通过三个支链的协调运动实现跟踪机构的高效跟踪。

1 新型并联跟踪机构原理和设计

1.1 机构原理

本文提出的并联跟踪机构如图1所示,由三个支链和上下平台组成,上平台为光伏板,下平台与地面固接。两个侧支链上端通过球铰或虎克铰与光伏板支架连接,下端通过铰链与下平台连接,中间支链上端的虎克铰通过移动副与光伏板支架连接,下端与下平台固接。

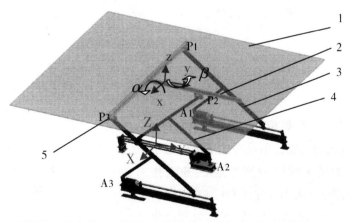

1-光伏板 2-光伏板T型支架 3-侧支链 4-中间支链 5-虎克铰

图1 并联跟踪机构

三个支链为曲柄滑块机构,利用丝杠螺母副实现滑块移动,带动连杆转动,实现光伏板运动。选取参考(定)坐标系XYZ固联于下平台,动坐标系xyz固联于光伏板T型支架上。三个支链滑块向相同方向移动,光伏板绕x轴转动;中间支链滑块不动,两侧支链滑块向相反方向运动,光伏板绕y轴转动;当中间支链和两侧支链滑块移动方向相反时,光伏板沿 z 轴上下平动,该运动不实现跟踪太阳的运动,当有大风天气时,通过该运动可降低光伏板平台高度,避免大风恶劣天气对跟踪机构装置的影响。两个侧支链结构相同,如图 2(a)所示为侧支链,图 2(b)所示为中间支链,中间支链和侧支链上铰接点 P 和 A 位于同一竖直线上。侧支链底端支架通过铰轴连接在下平台上,中间支链底端则与下平台固定,因此中间支链承受大部分重量,设计时需考虑底座强度。在三个支链的丝杠端安装编码器,用来检测三个滑块的运动位置,三个滑块的位置决定了光伏板的偏转角度和方向,根据滑块的距离和光伏板的角度以及太阳光照角度对应关系,在相应的时间点 PLC 系统读取数据库中电机的转速和起停命令,带动滑块移动到相应的位置,带动光伏板偏转到对应的角度。

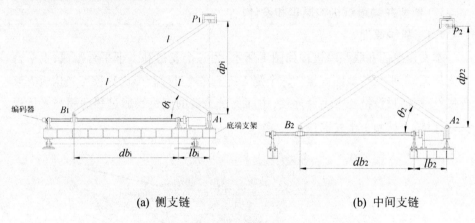

(a) 侧支链　　　　　　(b) 中间支链

图 2　支链结构简图

1.2　并联跟踪机构方位角与高度角设计

太阳高度角和方位角是在地平坐标系中描述的[9],如图 3 所示,O 点为地心,也看作观测点,观测点与太阳 S 连线 OS 与地平圈的角为太阳高度角 h,太阳的经度称为方位角 A,方位角是以南点为起点,沿地平圈向西度量。太阳高度角 h 和方位角 A 表示如下:

$$\sin h = \sin\varphi\sin\delta + \cos\varphi\cos\delta\cos t \quad (1)$$

$$\sin A = \frac{\cos\delta\sin t}{\cos h} \quad (2)$$

其中 δ 为太阳赤纬角,φ 为当地地理纬度,t 为太阳时角。并联太阳能跟踪机构是绕 x 轴

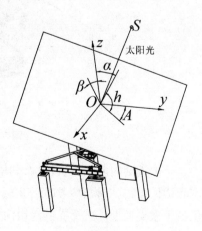

图 3　并联机构转动的角度 α, β

和 y 轴转动的,由图 3,可计算当太阳能高度角 h 和方位角 A 已知时,并联机构需绕 x 轴转动的角度 α 和绕 y 轴转动的角度 β。

$$\tan\alpha = \frac{\cos A}{\tan h}, \quad \tan\beta = \frac{\sin A}{\tan h} \tag{3}$$

根据兰州地区一年中太阳高度角和方位角变化最大范围值绘出图 4 所示太阳位置与机构转角关系图,从图中可以看出,并联跟踪机构转角范围 $-20° < \alpha < 85°$,$5° < \beta < 60°$。

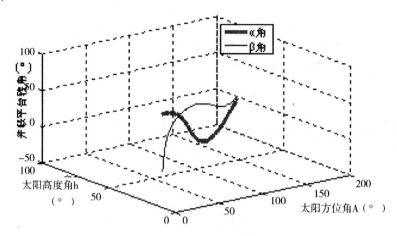

图 4　太阳位置与并联机构转角关系

2　运动分析

光伏板和底座的安装尺寸如图 5 所示,并联跟踪机构的运动学逆解是已知光伏板平台运动位置及方向角度 (α,β,z_t),z_t 为光伏板沿 z 方向运动的变量,求滑块的位移 d_{b1},d_{b2},d_{b3}。在光伏板中心建立动坐标系 cxyz,在底座中心建立定坐标系 OXYZ。

利用横滚(Roll)、俯仰(Pitch)、偏转(Yaw)角的姿态描述表示光伏板相对定坐标系的姿态为:

$$\begin{aligned} R_{py} &= Rot(Z,\gamma)Rot(Y,\beta)Rot(X,\alpha) \\ &= \begin{bmatrix} c\gamma c\beta & c\gamma s\beta s\alpha - s\lambda c\alpha & c\gamma s\beta c\alpha + s\gamma s\alpha \\ s\gamma c\beta & s\gamma s\beta s\alpha + c\gamma c\alpha & s\gamma s\beta c\alpha - c\gamma s\alpha \\ -s\beta & c\beta s\alpha & c\beta c\alpha \end{bmatrix} = \begin{bmatrix} n_x & o_x & a_x \\ n_y & o_y & a_y \\ n_z & o_z & a_z \end{bmatrix} \end{aligned} \tag{4}$$

式中 γ 表示并联机构绕 z 轴转动的角度,n_x、o_x、a_x、n_y、o_y、a_y、n_z、o_z、a_z 分别代表左矩阵中的对应项。

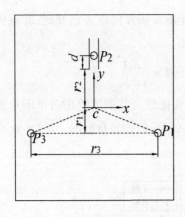 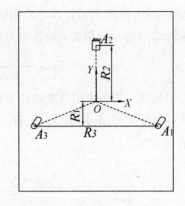

(a) 光伏板上平台　　　　　　(b) 下平台底座

图 5　上下平台安装尺寸

上平台相对定坐标系的位姿矩阵表示为

$$T_c^O = \begin{bmatrix} n_x & o_x & a_x & X_c \\ n_y & o_y & a_y & Y_c \\ n_z & o_z & a_z & Z_c \\ 0 & 0 & 0 & 1 \end{bmatrix} \quad (5)$$

式中 X_c、Y_c、Z_c 表示光伏板中心 C 相对定坐标系的坐标值。

上平台上点 $P_i(i=1,2,3)$ 相对动坐标系 $cxyz$ 齐次坐标表示为

$$\begin{aligned} P_1^c &= [r_3/2 \quad -r_1 \quad 0 \quad 1]^T \\ P_2^c &= [0 \quad r_2+d \quad 0 \quad 1]^T \\ P_3^c &= [-r_3/2 \quad -r_1 \quad 0 \quad 1]^T \end{aligned} \quad (6)$$

相对定坐标系 $OXYZ$ 表示为 $P_i^O = T_c^O P_i^c$,则有

$$\begin{aligned} P_1^0 &= \left[\frac{n_x r_3}{2} - o_x r_1 + X_c \quad \frac{n_y r_3}{2} - o_y r_1 + Y_c \quad \frac{n_z r_3}{2} - o_z r_1 + Z_c \quad 1\right]^T \\ P_2^0 &= [o_x(r_2+d)+X_c \quad o_y(r_2+d)+Y_c \quad o_z(r_2+d)+Z_c \quad 1]^T \\ P_3^0 &= [0 \quad 0 \quad 0 \quad 0]^T \end{aligned} \quad (7)$$

由图 5(a)可知,支链 1 上的 P_1 点在运动过程中保持与点 A_1 在同一竖直线上,因此有

$$\frac{n_y r_3}{2} - o_y r_1 + Y_c = -R_1 \quad (8)$$

同样 P_3 点在运动过程中保持与点 A_3 在同一竖直线上,有

$$-\frac{n_y r_3}{2} - o_y r_1 + Y_c = -R_1 \tag{9}$$

P_2 点相对于点 A_1 在 X 和 Y 方向不变,因此有

$$o_x(r_2 + d) + Y_c = 0, \; o_y(r_2 + d) + Y_c = R_2 \tag{10}$$

由式(7)-(10)可得光伏板中心 c 在定坐标系 $OXYZ$ 中位置和中间支链滑块位移

$$Y_c = -R_1 + o_y r_1 + \frac{n_y r_3}{2} = -R_1 + r_1 c\alpha$$

$$d = \frac{R_2 - Y_c}{o_y} - r_2 = \frac{R_2 - Y_c}{c\alpha} - r_2 \tag{11}$$

$$X_c = -o_x(r_2 + d) = -(r_2 + d)s\beta s\alpha$$

因此有

$$P_1^O = \left[\frac{r_3 c\beta}{2} - r_1 s\beta s\alpha + X_c \quad -r_1 c\alpha + Y_c \quad -\frac{r_3 s\beta}{2} - r_1 c\beta s\alpha + Z_c\right]^T$$

$$P_2^O = \left[(r_2 + d)s\beta s\alpha + X_c \quad (r_2 + d)c\alpha + Y_c \quad (r_2 + d)c\beta s\alpha + Z_c\right]^T \tag{12}$$

$$P_3^O = \left[-\frac{r_3 c\beta}{2} - r_1 s\beta s\alpha + X_c \quad -r_1 c\alpha + Y_c \quad -\frac{r_3 s\beta}{2} - r_1 c\beta s\alpha + Z_c\right]^T$$

点 A_1、A_2、A_3 在定坐标系 $OXYZ$ 中表示为

$$A_1^O = [R_3/2 \quad -R_1 \quad 0]^T$$

$$A_2^O = [0 \quad R_2 \quad 0]^T \tag{13}$$

$$A_3^O = [-R_3/2 \quad -R_1 \quad 0]^T$$

因此各支链滑块位移为

$$db_i = \sqrt{4l^2 - |P_1^O A_1^O|^2} - lb_i \tag{14}$$

3 结构力分析

并联跟踪机构运动由支链丝杠滑块上的驱动力 F_B 来驱动,支链结构受力分析如图 6 所示。

由图 6 得

$$\sum Y = F_B - F_{Ay} - F_{Py} = 0$$

$$\sum Z = F_{Bz} + F_{Az} - F_{Pz} = 0 \tag{14}$$

$$\sum M_A = F_{Py} l \sin\theta - F_{Bz} l \cos\theta = 0$$

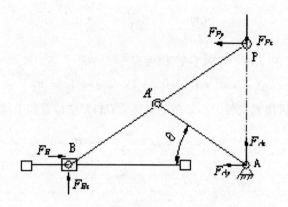

图 6　支链结构受力分析

有

$$F_{Bz} = F_{Py}\tan\theta$$
$$F_{Az} = F_{Pz} - F_{Bz} \quad (15)$$
$$F_{Ay} = F_B - F_{Py}$$

在支链机构系统中只有力 F_{Pz} 和力 F_B 作功,应用虚功原理有

$$\delta(2l\cos\theta)F_B + \delta(2l\sin\theta)F_{Pz} = 0$$

得

$$F_B = \frac{F_{Pz}}{\tan\theta} \quad (16)$$

由图 6 和式(16)可以看出,光伏板平台和机构的重量由 3 个支链共同支撑,每个支链上的重量由丝杠滑块上的驱动力 F_B 驱动,且驱动力随着角度 θ 的增大而减小。

4　结论

本文提出一种新型的并联太阳能跟踪机构,该机构运动范围大,承载能力强,由三个支链共同承载跟踪装置的重量。分析并得出了并联跟踪机构转角范围和太阳高度角和方位角关系,通过给出的机构运动学逆解表达式可计算出并联机构的杆件尺度。并联跟踪机构的结构受力分析得出各支链的驱动力表达式,为驱动电机的选型提供了依据。

参考文献

[1]吕文华,贺晓雷,于贺军,等. 全自动太阳跟踪器的研制和应用[J]. 光学精密工程,2008,16(12):2544-2550.

[2] Ardehali M M,Shahrestani M,adamsc. Energy simulation of solar assisted absorption sys-

tem and examination of clearness index effects on auxiliary heating［J］. Energy Conversion and Management,2007,78(3):864-870.

［3］刘俊,刘京诚,谢磊,等.太阳光自动跟踪装置的设计［J］.机床与液压,2010,38(9):45-48.

［4］张宝星.太阳能利用的跟踪与聚焦系统研究［D］.安徽:合肥工业大学,2006.

［5］黄真,赵永生,赵铁石.高等空间机构学［M］.北京:高等教育出版社,2006.

［6］刘善增,余跃庆,侣国宁,等.三自由度并联机器人的运动学与动力学分析［J］.机械工程学报,2009,45(8):11-17.

［7］张顺心,宋开峰,范顺成.基于并联球面机构的太阳跟踪装置研究［J］.河北工业大学学报,2003,32(6):44-48.

［8］贺新升,高春甫,王彬等.太阳自动跟踪机构的设计和位姿分析［J］.光学精密工程,2012,20(5):1048-1054.

［9］杨柯金.天球坐标系的难点解析［J］.南阳师范学院学报(自然科学版).2003,9:97-99.

注:本文曾发表在2015年《机械设计》第10期上

磁力电解复合抛光中带电粒子运动的分析

牛永江*

运用带电离子在电场中所受 Lorentz 力,建立了磁力电解复合抛光中带电粒子运动的数学模型,推导出了带电粒子的速度方程和轨迹方程,对三种典型带电粒子的运动进行了详细的比较和分析,通过对电磁场中带电粒子运动过程的分析,总结出了磁力电解复合抛光中磁场的作用。为了验证所建立的带电粒子模型的可靠性,对不锈钢材料进行了磁力电解抛光试验。试验结果和模型所分析的结果相符,证明了带电离子模型的合理性,以及对磁场作用分析的正确性。

1. 前言

对难加工金属材料而言,日本和中国的一些学者提出了复合抛光技术[1],即在传统电解抛光的基础上引入机械抛光和磁场的抛光工艺,如磁力曲面抛光、磁力机械电解抛光、磁力不织布电解抛光、磁力电子束抛光等。尤其针对不锈钢、SnSb 合金等难抛光材料,研究并提出了磁力电解复合抛光工艺[2]。但是,由于磁场对电解过程影响的复杂性,使得对磁场的作用,磁力电解复合抛光中磁场方向的设置,以及磁力电解复合抛光机理等问题的研究都极为困难,影响了磁力电解复合抛光工艺在工业领域的广泛使用。通过对磁力电解复合抛光中带电粒子运动情况的分析和研究,可以准确把握磁场对电解过程的影响情况及其作用,从而有助于对磁力电解复合抛光机理的研究,奠定磁力电解复合抛光的理论基础[3][4]。

---

* 作者简介:牛永江(1965—),男,甘肃天水人,天水师范学院教授、学士,主要从事特种加工制造技术研究。

2. 带电粒子的运动分析

2.1 带电粒子运动方程的建立

磁力电解复合抛光是在电解抛光的基础上,两电极间附加一磁场而形成的复合抛光工艺。带电粒子在电场与磁场的共同作用下运动[5-7],如图1所示。作用力F

$$F = q(E + v \times B) \tag{1}$$

式中　F——作用力
　　　　q——带电粒子电量
　　　　E——电极间电场强度
　　　　V——带电粒子运动速度
　　　　B——磁场强度

$$E = E e_y \tag{2}$$

$$v = \left(\frac{dx}{dt}\right)e_x + \left(\frac{dy}{dt}\right)e_y + \left(\frac{dz}{dt}\right)e_z \tag{3}$$

e_x, e_y, e_z 为单位方向矢量。

在磁力电解复合抛光中,通常外加和电场垂直的磁场

$$B = B e_z \tag{4}$$

$$m\left(\frac{dv}{dt}\right) = F \tag{5}$$

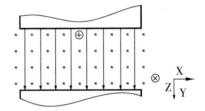

图1　电磁场中的带电粒子

将(1)—(4)式带入(5)式,解得:

$$\begin{cases} \dfrac{dx}{dt} = C_1 \cos(\omega t + \alpha) + \dfrac{E}{B} \\ \dfrac{dy}{dt} = -C_1 \sin(\omega t + \alpha) \\ \dfrac{dz}{dt} = C_2 \end{cases} \tag{6}$$

$$\begin{cases} x = \dfrac{C_1}{\omega} \sin(\omega t + \alpha) + \dfrac{E}{B} t + x_0 \\ y = \dfrac{C_1}{\omega} \cos(\omega t + \alpha) + y_0 \\ z = C_2 t + z_0 \end{cases} \tag{7}$$

式中 $\omega = \dfrac{qE}{m}$, $C_1, C_2, \alpha, x_0, y_0, z_0$ 为积分常数。对于电解液中的带电粒子,设 t = 0

时，带电粒子的轨迹方程为：

$$\begin{cases} x = \dfrac{m\sqrt{\left(\dfrac{E}{B}-v_{x0}\right)^2+v_{y0}^2}}{qE}\sin\left(\dfrac{qE}{B}t+\alpha\right)+\dfrac{E}{B}t+\dfrac{mv_{y0}}{qE}+x_0 \\ y = \dfrac{m\sqrt{\left(\dfrac{E}{B}-v_{x0}\right)^2+v_{y0}^2}}{qE}\cos\left(\dfrac{qE}{B}t+\alpha\right)-\dfrac{mv_{x0}}{qE}+\dfrac{m}{qB}+y_0 \\ z = v_{z0}t+z_0 \end{cases} \quad (8)$$

2.2 带电粒子运动分析

磁力电解复合抛光时，两极间的带电粒子，首先，受到电场力和洛伦磁力作用在两极间做复杂运动；其次，又有带电粒子相互间碰撞和随机作用引起的四散运动。因此，造成了整个运动的复杂性和随机性，要准确分析每个带电粒子的运动极其困难，为了认识和掌握磁力电解复合抛光中带电粒子运动，尤其磁场对带电粒子的影响，我们暂不考虑离子间相互的碰撞作用，并且只对 $x_0=0,y_0=0,z_0=0$ 时三种典型带电粒子的运动加以分析[8][9]。

1) $v_{x0}=v_0,v_{y0}=0,v_{z0}=0$ 的带电粒子

将 $v_{x0}=v_0,v_{y0}=0,v_{z0}=0$ 代入(8)式，得出磁力电解复合抛光中该类带电粒子的轨迹方程为：

$$\begin{cases} x = \dfrac{v_0-v_E}{\omega}\sin\omega t+v_E t \\ y = \dfrac{v_0-v_E}{\omega}(1-\cos\omega t) \\ z = 0 \end{cases} \quad (9)$$

式中 $v_E=\dfrac{E}{B}$，从式(9)看出，带电粒子发生了电漂移。若 $v_0<\dfrac{E}{B}$，则带电粒子的轨迹时短幅摆线，当 v_0 增大时，离子轨迹向 y 轴靠近。若 $v_0=\dfrac{E}{B}$，带电粒子轨道沿 y 轴做匀速直线运动。

2) $v_{x0}=0,v_{y0}=v_0,v_{z0}=0$ 的带电粒子

将 $v_{x0}=0,v_{y0}=v_0,v_{z0}=0$ 代入(8)式，得出磁力电解复合抛光中该类带电粒子的轨迹方程为：

$$\begin{cases} x = \dfrac{\sqrt{v_E{}^2 + v_{0y}{}^2}}{\omega}\sin(\omega t + \alpha) + v_E t + \dfrac{v_{0y}}{\omega} \\ y = \dfrac{\sqrt{v_E{}^2 + v_{0y}{}^2}}{\omega}\cos(\omega t + \alpha) + \omega \\ z = 0 \end{cases} \quad (10)$$

从式(10)可以看到,该类带电粒子以半径为 $\dfrac{\sqrt{v_E{}^2 + v_{0y}{}^2}}{\omega}$,圆心为($v_E t + \dfrac{v_{0y}}{\omega}$, ω)的变心圆周运动,其运动频率为 ω 。 v_E 和 v_{0y} 越小,圆周半径越小,即偏转越大; v_E 和 v_{0y} 越大,圆周半径越大,即偏转越小。

3) $v_{x0} = 0, v_{y0} = 0, v_{z0} = v_0$ 的带电粒子

将 $v_{x0} = 0, v_{y0} = 0, v_{z0} = v_0$ 代入(8)式,得出磁力电解复合抛光中该类带电粒子的轨迹方程为:

$$\begin{cases} x = \dfrac{v_E}{\omega}\sin\omega t + v_E t \\ y = \dfrac{v_E}{\omega}\cos\omega t + \omega \\ z = v_0 t \end{cases} \quad (11)$$

从式(11)可以看到,该类带电粒子的轨迹为变圆心螺旋线运动,在 XY 平面内作变圆心运动,半径为 $\dfrac{v_E}{\omega}$,圆心为($v_E t$, ω),沿 Z 轴作匀速运动。 v_E 越大,圆周半径越大。

2.3 离子运动方程的比较及磁场作用分析

如果没有外加磁场,只有电场作用时,带电粒子只受到电场力的作用 F,其轨迹方程为:

$$\begin{cases} x = v_{x0} t + x_0 \\ y = \dfrac{1}{2} qE t^2 + v_{y0} t + y_0 \\ z = v_{z0} t + z_0 \end{cases} \quad (12)$$

将 $v_{x0} = v_0, v_{y0} = 0, v_{z0} = 0$ 、 $v_{x0} = 0, v_{y0} = v_0, v_{z0} = 0$ 和 $v_{x0} = 0, v_{y0} = 0, v_{z0} = v_0$ 分别代入式(12),得出 $v_{x0} = v_0, v_{y0} = 0, v_{z0} = 0$ 、 $v_{x0} = 0, v_{y0} = v_0, v_{z0} = 0$ 和 $v_{x0} = 0, v_{y0} = 0, v_{z0} = v_0$ 这三类带电粒子在无外加电场时的轨迹方程,分别如下:

$$\begin{cases} x = v_0 t \\ y = \dfrac{1}{2} qEt^2 \\ z = 0 \end{cases} \tag{13}$$

$$\begin{cases} x = 0 \\ y = \dfrac{1}{2} qEt^2 + v_0 t \\ z = 0 \end{cases} \tag{14}$$

$$\begin{cases} x = 0 \\ y = \dfrac{1}{2} qEt^2 \\ z = v_0 t \end{cases} \tag{15}$$

通过对三类典型带电粒子在有外加磁场的轨迹方程与无外加磁场的轨迹方程的比较,可以看到,磁场在电解过程作用如下:

1)磁场能显著提高电解抛光的速度。由于外加磁场的作用,使得电极间 $v_{x0} = v_0, v_{y0} = 0, v_{z0} = 0$ 类带电粒子发生了电漂移;同时,由于外加磁场的作用,使得 $v_{x0} = 0, v_{y0} = v_0, v_{z0} = 0$ 类带电粒子的运动轨迹发生了偏转,而这两种因素都降低了电解过程中的浓差极化,从而提高了电解抛光的速度。

2)磁场能显著提高电解抛光的效率。由于外加磁场的作用,使得 $v_{x0} = 0, v_{y0} = v_0, v_{z0} = 0$ 和 $v_{x0} = 0, v_{y0} = 0, v_{z0} = v_0$ 两类带电离子到达电极的概率增大,提高了电解反应的效率,从而提高了电解抛光的效率。

3)磁场能显著提高被抛光材料的表面质量。磁场像一个无形的搅拌器,使得三种典型带电离子的运动轨迹发生了巨大变化,电极间带电粒子的的运动复杂化、反应过程均匀化,提高了被抛光材料的表面质量。

3 磁力电解复合抛光试验
3.1 试验条件

1)试验条件:

电解液:20% $NaNO_3$,

磁感应强度:0.1T

电流密度:0.15 – 0.25 A/cm^2

主轴转速:17500rpm

被抛光前工件表面粗糙度为 Ra = 3.2μm。

2）试验原理[10-12]：

在数控机床主轴上安装一个球型磨头,磁力电解复合抛光片被固定在磨头上。磨头和工件之间通入电解液。工件接电源正极,磨头接电源负极。工件和磨头之间加一个和电场正交的磁场,其实验结构原理图2所示[7][8]。

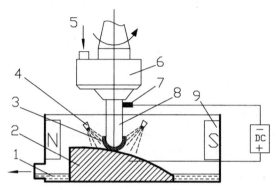

1 电解液 2 工件 3 磁力电解复合抛光片 4 喷嘴
5 冷却液 6 增速装置 7 电刷 8 磨头 9 磁铁

图2 实验结构原理图

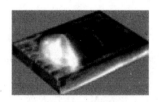
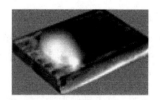

a) 无外加磁场　　　　b) 有外加磁场

图3 有无磁场时抛光后工件表面图

3.2 试验结果

在以上实验条件下,对如图3所示不锈钢工件抛光,被抛光前不锈钢工件表面粗糙度 $Ra=3.2\mu m$,抛光30min后,无磁场时,表面粗糙度 $Ra=1.2\mu m$,引入磁场后,表面粗糙度 $Ra=0.2\mu m$。

4 结论

通过对磁力电解复合抛光中带电粒子运动的分析,以及实验对磁力电解复合抛光过程的验证，可以看出：

1）磁力电解复合抛光是一种抛光速度更快、抛光效率更高和抛光后表面质量更好的抛光工艺，能够广泛用于难加工金属材料的精加工中。

2）通过对 $v_{x0} = v_0, v_{y0} = 0, v_{z0} = 0$、$v_{x0} = 0, v_{y0} = v_0, v_{z0} = 0$ 和 $v_{x0} = 0, v_{y0} = 0, v_{z0} = v_0$ 三种典型带电粒子运动的分析，得出的磁场在磁力电解复合抛光中的作用是正确的。

3）带电粒子运动的数学模型是合理的，速度方程和轨迹方程是正确的。

参考文献

[1] Swain, John. Where to Use Electropolishing. Design Engineering, 1999, (9): 47 – 48.

[2] Chen Yuquan. Study of the Compound Polishing Process on SnSb alloy Moulds[J]. Journal of Materials Processing Techology, 2002, (129): 310 – 314.

[3] Hon – yuen Tam, Osm and Chi – hang Lui, Alberert CK MOK. Robotics Polishing of Free – form Surface Using Scanning Paths. Journal of Materials Processing Technology, 1999, (95): 191 – 200

[4] J. P. Huissoon, F. Ismail. Automated Polishing of Die Steel Surfaces. The International Journal of Advanced Manufacturing Technology, 2002. 19(4): 285 – 29

[5] 刘晋春, 赵家齐主编. 特种加工[M]. 机械工业出版社, 1998年5月第2版.

[6] 刘爱华, 徐中耀, 邓兴华. 碳钢——不锈钢复合金属电化学抛光工艺研究[J]. 电加工, 1998, (2): 34 – 36.

[7] Kim Jeong – Du, Jin Dong – Xie, Choi Min – Seng. Study on the Effect of a Magnetic Field on an Electrolytic Finishing Process. International Journal of Machine Tools & Manufacture, 1997, 37(4)

[8] [日]微细加工技术编辑委员会编. 微细加工技术[M]. 北京: 科学出版社, 1983

[9] Chen Yuquan. The Effect of the Magnetic – Field on the Electrolytic – Abrasive Finishing [C]. ICPCG – 1996, 1996.

[10] 张永俊, 刘晓宁. 机械电解复合抛磨曲面的工具设计[J]. 广东工业大学学报, 2000, 17(3): 22 – 25.

[11] 彭敏, 曲宁松, 朱荻. 不锈钢电解抛光工艺研究[J]. 航空精密制造技术, 2001(3): 6 – 10.

[12] E. – S. Lee. Machining Characteristics of the Electropolishing of Stainless Steel (STS316L). Advanced manufacturing technology, 2000, 16(8): 591 – 599

注: 本文曾发表在2010年《天水师范学院学报》第2期上

工件定位基准的定义及其确定新探

牛永江[*]

分析了工件轮廓要素和中心要素之间的关系，提出了应统一以直接与定位元件的定位表面接触的工件上的轮廓要素为定位基准的观点。

要计算定位误差，就必然涉及到定位基准的确定，目前一般的夹具设计论著中，都将定位基准分为工件上直接与定位元件接触的轮廓要素和没有与定位元件接触的中心要素，如孔的中心线、轴的中心线和对称中心平面等。这种划分方式容易引起概念混淆，使定位基准的判定复杂化，在某些定位状态下，还会造成定位基准不重合，导致定位误差计算烦琐困难。

定位基准是工件定位时，代表工件在夹具中所占位置的点、线、面等几何要素[1]，对于一个工件的轮廓要素和中心要素，从相互关系方面分析，两者之间是有差异的，但两者同时又是统一的，是同一个形体上的几何要素，互相反映、互相体现，都可以代表工件，所以在判断定位基准时，不应该撇开直接与定位元件接触具体完成定位的轮廓要素，而非要把由轮廓要素所间接体现的中心要素确定为定位基准，基于此，笔者认为在计算定位误差时，应统一以直接与定位元件的定位表面接触的工件上的轮廓要素为定位基准，如孔和轴的母线等。这样定义：

其一，使定位基准的内涵更清晰明了，定义更准确精炼，既避免了概念歧义的产生，又防止了认识上的混淆；

其二，有效简化了定位基准的判定过程，使得定位基准的确定更加直观准确。

那么，统一以工件的轮廓要素作为定位基准，是否会影响定位误差的计算结

---

[*] 作者简介：牛永江（1965— ），男，甘肃天水人，天水师范学院教授、学士，主要从事特种加工制造技术研究。

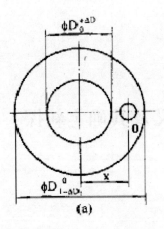 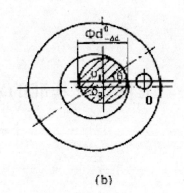

图1

果,下面试通过两种典型的以中心要素为定位基准的定位方案加以说明,设定位误差用 Δ 表示,基准不重合误差用 Δ_B 表示,基准位移误差用 Δ_Y 表示。

例一:在图1(a)所示工件上钻一孔 D,定位方案如图 1(b)所示,求工序尺寸 x 的定位误差,设最小安装间隙为 0。[2]

解:工件以内圆柱面在心轴上定位,当以两者的接触母线为定位基准,则:

因为 $\Delta_X = \Delta_Y + \Delta_B$,$\Delta_Y = \Delta d/2\ cos\theta$,$\Delta_B = \Delta D/2\ cos\theta$,

所以 $\Delta_X = \Delta d/2\ cos\theta + \Delta D/2cos\theta = (\Delta d + \Delta D)/2\ cos\theta$;

当按传统方法,以孔的回转轴线为定位基准,则:

因为 $\Delta_X = \Delta_Y + \Delta_B$,$\Delta_Y = O_1O_2\ cos\theta = (\Delta d + \Delta D)/2\ cos\theta$,$\Delta_B = 0$,

所以 $\Delta_X = (\Delta d + \Delta D)/2\ cos\theta + 0 = (\Delta d + \Delta D)/2\ cos\theta$;

例二:如图2所示,工件以外圆柱面在 V 形块上定位铣键槽,工件外径为 D ± ($\delta_d/2$),求工序尺寸 H 的定位误差,设 V 形块的夹角为 α。[3]

解:工件以外圆柱面在在 V 形块上定位,当以两者的接触母线为定位基准,则:

因为 $\Delta_H = \Delta_Y - \Delta_B$,

Δ_B = BC - B'C' = (OA - O'A') - (OB - O'B')

$= \delta_d/2 - [OA\ sin(\alpha/2) - O'A'sin(\alpha/2)]$

$= \delta_d/2 - (OA - O'A')sin(\alpha/2)$

$= \delta_d/2\ [1 - sin(\alpha/2)]$;

Δ_Y = BB' = (AB - A'B')ctg($\alpha/2$)

= (OA - O'A')cos($\alpha/2$)ctg($\alpha/2$)

= $\delta_d/2\ cos(\alpha/2)ctg(\alpha/2)$;

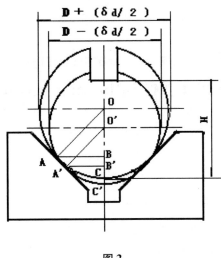

图2

所以 $\Delta_H = \delta_d/2 \cos(\alpha/2) \operatorname{ctg}(\alpha/2) \delta_d/2 \cos(\alpha/2)$
  $= \delta_d/2 [1/\sin(\alpha/2) - 1]$；

当按传统方法,以外圆柱面的中心线为定位基准,则：

因为 $\Delta_H = \Delta_Y - \Delta_B, \Delta_B = \delta_d/2$,

$\Delta_Y = OO' = (D + \delta_d/2)/2\sin(\alpha/2) - (D - \delta_d/2)/2\sin(\alpha/2) = \delta_d/2\sin(\alpha/2)$,

所以 $\Delta_H = \delta_d/2\sin(\alpha/2) - \delta_d/2 = \delta_d/2 [1/\sin(\alpha/2) - 1]$；

由以上计算结果可以看出,以轮廓要素(接触母线)为定位基准和以传统的中心要素(回转轴线)为定位基准,其定位误差的计算结果是相同的。因此,完全可以统一以直接与定位元件的定位表面接触的工件上的轮廓要素作为定位基准。

参考文献

[1] 李庆寿. 机床夹具设计[M]. 北京：机械工业出版社,1984.17.

[2] 秦国华. 定位误差计算模型的建立及应用[J]. 工具技术,1999,33(8):30 - 34.

[3] 东北重型机械学院. 机床夹具设计手册[M]. 上海：上海科学技术出版社,1990. 6 - 29.

注：本文曾发表在2003年《机械研究及应用》第1期上

Static and dynamic characteristics modeling for CK61125 CNC lathe bed basing on FEM

Hongping Yang*

摘 要:机床床身特性直接影响机床的加工精度。首先,基于有限元分析方法,对 CK61125 数控车床进行了静态和动态结构建模。结果表明,最大应力为 100.98MPa,最大应变为 0.615μm/mm,最大变形为 0.1455m m。其次,基于有限元分析方法对模态进行了分析,结果表明:一阶共振频率为 271.63Hz,二阶共振频率为 290.41Hz,三阶共振频率为 305.88Hz。

The lathe bed characteristics are directly effect on the machining accuracy. Firstly, based on finite element analysis method, the static and dynamic structural are modeled for CK61125 CNC lathe. The results show that the maximum stress is 100.98MPa, the maximum strain is 0.615μm/mm, the maximum deformation is 0.1455mm. Secondly, the modal is analyzed based finite element analysis method, the results show that the first - order resonance frequency is 271.63Hz, the second - order resonance frequency is 290.41 Hz, the third - resonance frequency is 305.88Hz.

1. Introduction

Machine tool is the main manufacturing equipment, and the computer numerical control(CNC)machine tool has high - precision, high - efficiency automated production equipment. It is widely used in automotive, aviation, aerospace, marine, rail, wind, hydroelectric power and other fields. With the growing demand for complex precision parts, it made the CNC machine tools preferment development to efficiency, reliability, precision level developed[1-2].

---

* 作者简介:杨红平(1974—),男,甘肃秦安人,天水师范学院教授、博士(后),主要从事机电系统非线性动力学研究。

The lathe bed dynamic and static performances have an important impact on the performance of the machine. The structural characteristics of the bed are related to the machine's machining accuracy and quality, as well as the machine's operating stability. So the designer must study the bed dynamic and static performance.

With the development of computer technology and the analysis method, the digital manufacturing is realized graduallyto information technology and manufacturing technology[3-6]. Among those methods, the finite element analysis method is an advanced and effective design method because the finite element analysis method can get the stress, distribution, deformation, stress easily and accurately. And the internal stress of components can be calculated. This makes the design of structure more targeted. Tyalor[7] based the theory of quality, drilling machine has carried basing the finite element modeling. The analysis result is reliable. However, due to ignoring the influence of joint stiffness and damping coefficient, the dynamic characteristics of machine tool was not true. Jiang[8-9] proposed mathematical model to simulate the machine tool connection structure method using finite element analysis. Machine tool spindle to the position and number of connection of CNC machine tools is optimized. Yeh[10-11] established the modal for movable milling machine bed machine tool bed, head frame and column used finite element method. And modal analysis was done for each part. He[12] modeled the solid model of CA6140 machine bed by using UG software, then the solid model was imported into the ANSYS software in order to process its modal analysis. The spacing, thickness and structure of the ribs were looked as the design variables. The correspondence was established between the bed natural frequency and stiffness, and got the optimal results. Wang[13], based on the dynamic optimization principle of structure and the variable analysis method of finite element method, the dynamic characteristics of the bed structure of precision machine tools are analyzed. Two structural optimization schemes for the bed structure are proposed basing the results of the meta-structure analysis. Abuthakeer[14] is to improve the stiffness, natural frequency and damping capability of machine tool bed using a composite material containing welded steel and polymer concrete. A machine tool bed made of sandwich structures and polymer concrete are designed and manufactured. Modal and static analyses were conducted numerically and experimentally to determine the modal frequencies, damping ratio, deformation and strain.

In this paper, as the research object of CK61125 horizontal CNC lathe, the CNC lathe bed unit are modeled and analyzed based the finite element analysis method in analyzing structure and working characteristics of the lathe bed deeply. The lathe bed working stress, strain and the modal parameters are calculated. The static, dynamic characteristics analysis principle

The static and dynamic finite element modeling of machine lathe bed is set up the finite element digital model, which can reflect the engineering practice through the analysis of its performance. The basic ideas are: to discrete the continuous structure into a finite number of units, and set a finite number of nodes in each unit, to make continuum as only in the node connected to a set of assembly unit, to select the node value as the basic unknown quantity, to assume an approximate interpolation function in each unit to represent the unit midfielder function distribution, using the mechanics principle such as variation principle and virtual work principle to solve the unknown node finite element equation, which will be a continuous domain of infinite degrees of freedom into the limited degree of freedom of discrete domain problems.

1.1 Static characteristic analysis

Used the principle of elasticity mechanics, elastic body under the load of external force, any point the stress, strain, deformation respectively are:

$$\{\sigma\} = [\sigma_x \ \sigma_y \ \sigma_z \ \tau_{xy} \ \tau_{yz} \ \tau_{zx}]^T \tag{1}$$

$$\{\delta\} = [u \ v \ w]^T \tag{2}$$

$$\{\varepsilon\} = [\varepsilon_x \ \varepsilon_y \ \varepsilon_z \ \gamma_{xy} \ \gamma_{yz} \ \gamma_{zx}]^T \tag{3}$$

For isotropic linear elastic material, the relationship between the stress and strain vector is:

$$\{\sigma\} = [D]\{\varepsilon\} \tag{4}$$

where, $[D]$ is the elasticity matrix, its value depends entirely on the lathe bed materials elastic modulus E and poisson's ratio μ, its expression is:

$$[D] = \frac{E(1-\mu)}{(1+\mu)(1-2\mu)} \begin{bmatrix} 1 & \frac{\mu}{1+\mu} & \frac{\mu}{1-\mu} & 0 & 0 & 0 \\ & 1 & \frac{\mu}{1-\mu} & 0 & 0 & 0 \\ & & 1 & 0 & 0 & 0 \\ & & & \frac{1-2\mu}{2(1-\mu)} & 0 & 0 \\ & & & & \frac{1-2\mu}{2(1-\mu)} & 0 \\ & & & & & \frac{1-2\mu}{2(1-\mu)} \end{bmatrix} \tag{5}$$

Using the virtual displacement principle, the external force to the work done on the virtual displacement is equal to the virtual strain on the work done corresponding to the stress with the virtual displacement. Assuming the virtual strain matrix is $\{B\}$, the relationship between the finite element virtual strain $\{\varepsilon\}^*$ and the virtual displacement $\{\delta\}^*$ are:

$$\int_e \{\delta\}^{*T}[B]^T[D][B]\mathrm{d}V = \{\delta\}^{*T}\{F\}^e \tag{6}$$

$$\int_e \{\varepsilon\}^{*T}\{\sigma\}\mathrm{d}V = \{\delta\}^{*T}\{F\}^e \tag{7}$$

$$\int_e [B]^T[D][B]\mathrm{d}V\{\delta\}^e = \{\delta\}^{*T}\{F\}^e \tag{8}$$

$$[K]^e\{\delta\}^e = \{F\}^e \tag{9}$$

$$[K]^e = \int_e [B]^T[D][B]\mathrm{d}V \tag{10}$$

To calculate each element node corresponding to the structure load matrix and stiffness matrix, it can get the balance equation of integral node.

$$[K]\{\delta\} = \{F\} \tag{11}$$

Through the external displacement boundary conditions, the structure node displacement $\{\delta\}$ can be obtained. And then, with a known element node displacement, the corresponding stress and the strain value of the unit can calculated.

1.2 Dynamic characteristic analysis

Modal analysis is the basis of dynamic analysis. The modal can be obtained by analyzing the system inherent frequency, vibration mode, natural frequency and vibration mode. If the elastic system have multiple freedoms, we can derived its motion equation using the dynamic loading and virtual work principle. Then we can get the spindle dynamic equilibrium equation by the element strength, surface force, concentrated force vector $\{F(t)\}$:

$$[M]\{\ddot{x}(t)\} + [C]\{\dot{x}(t)\} + [K]\{x(t)\} = \{F(t)\} \tag{12}$$

Where, $\{x(t)\}$ is the spindle structure displacement, $\{\dot{x}(t)\}$ is the lathe bed structure total velocity, $\{\ddot{x}(t)\}$ is the lathe bed structure total acceleration. The lathe bed stiffness matrix, mass matrix and damping matrix are $[K]$, $[M]$ and $[C]$, respectively.

In the process of modal analysis, taking the $\{F(t)\}$ is the zero matrix. Because the structural damping is small, the structure of the natural frequency and mode of influence is negligible, which can be obtained by the structure of undamped free vibration equation:

$$[M]\{\ddot{x}(t)\} + [K]\{x(t)\} = 0 \tag{13}$$

In free vibration, each mass particle is in situation of simple harmonic vibration. if $\{x_0(t)\}$ is the node amplitude matrix, ω is the frequency corresponding to vibration mode. Then, each node displacement can be expressed as:

$$\{x(t)\} = \{x_0(t)\}\cos(\omega t) \tag{14}$$

Hence, the homogeneous equation can be obtained as:

$$([K] - \omega^2[M])\{x_0(t)\} = 0 \tag{15}$$

When lathe bed is in free vibration, the amplitude of all each node is not zero, the coefficient determinant must be zero. So the lathe bed modal frequency equation is obtained:

$$\left|[K] - \omega^2[M]\right| = 0 \tag{16}$$

So, the lathe bed each order of the natural vibration frequency and main modes can be obtained.

2. The lathe bed structure parameters

Taking CK61125 horizontal CNC lathe bed as the research object, the bed structures is shown inFig. 1. The lathe bed material is HT250. Its elastic modulus is 140 GPa, poison's ratio is 0.25, the material density is 7890 kg · m^{-3}, the yield strength is 240 MPa.

The lathe bed is loaded by the spindle system, cutting forces et al. Its loading is shown inFig. 1. The pressure value of B, D, E, F, G, H, I surface is 0.0139MPa, 0.022MPa, 0.0218MPa, 0.0305MPa, 0.0253MPa, 0.0428MPa, 0.0428MPa, respectively. The moment value of C, J surface is 14.9kN · m, 2.68kN · m, respectively.

3. The lathe bed static, dynamic characteristics modeling and analysis

3.1 The static characteristic modeling results analysis

According to the situation of processing machine tools, the lathe bed stress and strain are analyzed, as show in Fig. 2.

From Fig. 2(a), it can be seen that the lathe bed maximum stress is 80.49 MPa.

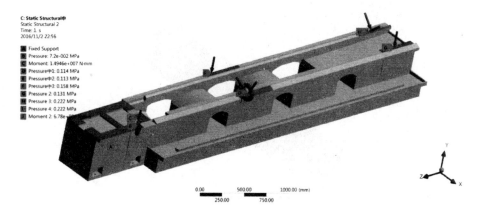

Fig. 1. The lathe bed loadings.

However, the HT250 allowable stress is 120 MPa. The calculated value is less than the allowable stress. From Fig. 2(b), it can be seen that the maximum strain is 0.6147μm /mm. By Fig. 2(c), we can see, the bed biggest deformation is 0.1455 mm under the loading.

3.2 The lathe bed modal analysis

Modal is inherent vibration characteristics of the object structure. It is one of the main parameters for object in the design of dynamic loading structure. Each order modal has its different natural frequency and modal vibration mode. Each order modal analysis of the results is an important basis for structure design, and it can avoid harm under the action of dynamic load resonance. At the same time, vibration modal can judge the mechanical parts weak in the working, and it can optimize the mechanical parts design. So the modal analysis is an important content in the design of machine tool.

This paper present the lathe bed modal calculate results using subspace method. Its four order naturalfrequencies and vibration mode are obtained, as shown in Table 1 and Fig. 3.

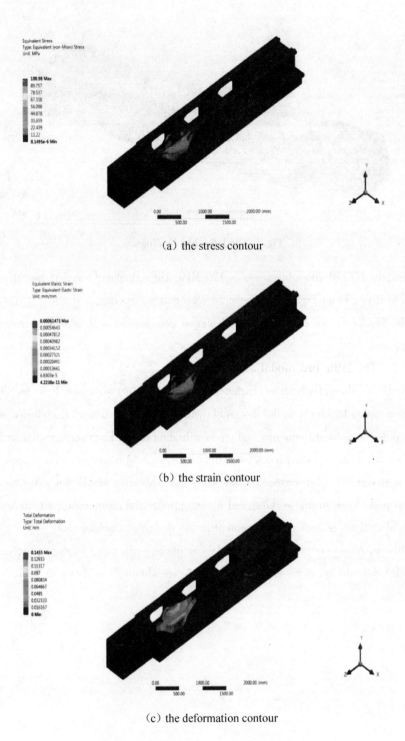

(a) the stress contour

(b) the strain contour

(c) the deformation contour

Fig. 2 The lathe bed calculating contour.

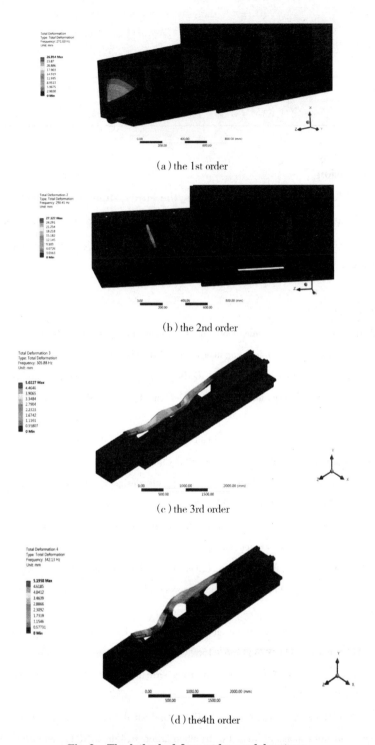

(a) the 1st order

(b) the 2nd order

(c) the 3rd order

(d) the 4th order

Fig. 3　The lathe bed four order modalcontour.

Table 1 The results of bed modal.

Modal order	The 1st order	The 2nd order	The 3rd order	The 4th order
Frequency(Hz)	271.63	290.41	305.88	342.13

4. Conclusion

For the CK61125 CNC lathe bed parts, the stress, strain, deformation and modal are analyzed basing FEM, the stress and strain has been shown on the lathe bed structure parameters and the modal frequency. It will provide strong theoretical parameters for the machine tool, and modeling analysis methods.

References

[1] H. R. Cao, Z. J. He, Dynamics modeling and model updating system between machine tool and its spindle, Journal of mechanical engineering, 13(2012)88 - 94.

[2] V. Gagnol, B. C. Bouzgarrou, P. Ray, C. Barra, Model - based chatter stability prediction for high - speed spindles, International Journal of Mechine Tools and Manufacture, 47(2007)1176 - 1186.

[3] Y. L. Jia, T2120 deep hole machine tool spindle static and dynamic characteristics analysis and optimization design, Taiyuan, North University of China, 2015.

[4] O. Maeda, Y. Z. Cao, Y. Altintas, Expert spindle design system, International Journal of Machine Tools & Manufacture, 3(2005)537 - 548.

[5] A. Tedric, N. Michael, Rolling bearing analysis fifth edition, advanced concepts of bearing technology, New York: Taylor & Francis Group, 2006.

[6] A. A. Mohammed, A. A. Esharkawy, Effect of axial preloading of angular contact ball bearings on the dynamics of a grading machine spindle system, Journal of Materials Processing Technology, 136(2003)48 - 59.

[7] M. P. Rossow, E. Tyalor, A finite element method for the optimal design of variable thickness sheets, AIAA Journal, 11(1973)1566 - 1569.

[8] T. Jiang, M. Ciredast, Virtual assembly using virtual reality techniques, Computer - Aided Design, 29(1997)575 - 584.

[9] B. Lia, J. Hong, Z. Wang, W. Wu, Y. Chen, Optimal design of machine tool bed by load bearing topology identification with weight distribution criterion, 45th CIRP Conference on Manufacturing Systems, Procedia CIRP, 3(2012)626 - 631.

[10] T. P. Yeh, J. M. Vance, Applying virtual reality techniques to sensitivity based structural shape design, Journal of Mechanical Design, Transaction of the ASME, 120(1998)619-621.

[11] M. Mori, M. Fujishima, Y. Inamasu, Y. Oda, A study on energy efficiency improvement for machine tools, Manufacturing Technology, 60(2011)145-148.

[12] H. J. He, J. Mei, L. X. Ma, Design and optimization of machine bed CA6140, Journal of Chongqing University of Technology(Natural Science), 29(2015)1674-8425.

[13] F. Q. Wang, Z. Y. Rui, C. L. Lei, Q. Wu, Machine tool bed based on unit structure Structural dynamic analysis and optimization of precision, Journal of Machine Design, 29(2012)93-96.

[14] S. S. Abuthakeer, P. V. Mohanram, G. Mohankumar, Static and dynamic performance improvement of conventional computer numerical control machine tool bed with hybrid welded steel, American Journal of Applied Sciences, 8(2011)610-616.

注:本文曾发表在2017年《Procedia Engineering》174卷上

基于分形几何与接触力学理论的结合面法向接触刚度计算模型

杨红平*

基于分形几何理论和接触力学理论,用分形理论表征粗糙表面微凸体参数,考虑微凸体由弹性变形向弹塑性变形以至最终向完全塑性变形转化的过程,建立了各变形阶段微凸体的接触刚度模型。在此基础上,提出了机械结合面法向接触刚度计算模型,该模型揭示了在不同的塑性指数下,结合面法向接触载荷与法向接触刚度之间的关系。结果表明,在塑性指数较小时,微凸体的变形以弹性为主,法向接触载荷与接触刚度之间表现为近似线性关系;随着塑性指数的增加,微凸体变形主要以塑性为主,法向接触载荷与接触刚度之间表现为较强非线性关系。对已有的铣削加工和磨削加工情况下的结合面法向接触刚度实验结果,利用该模型进行数值计算、仿真和分析。结果表明:本文模型较 JZZ 模型相比更与实验曲线吻合。

0 前言

机床结构存在着大量相互接触的机械结合面。研究表明,机床结合面的接触刚度约占机床总刚度的 60%–80%[1]。由此可见,机床结合面刚度(包括法向刚度和切向刚度)对机床整机产品的性能具有非常重要的影响。方便、准确地获取可靠的结合面刚度参数,是机床包括高档数控机床乃至一般机械产品的研发、分析与设计过程中存在的一个关键理论与技术问题,是国内外学术界研究热点。

本文研究结合面法向接触刚度的理论计算模型。传统的方法大多是是通过实验直接获得,或者通过参数识别间接获得在一定条件下的结合面法向刚度。影响结合面刚度的因素主要包括结合面的材质、表面加工方法、表面粗糙度、结合面

---

* 作者简介:杨红平(1974—),男,甘肃秦安,天水师范学院机教授,博士(后),主要从事机电系统非线性动力学研究。

面压、接触面间介质等,这些因素之间具有强的非线性特性[2-8],黄玉美、傅卫平等[3-6]利用实验获得了结合面的法向特性参数。但实验数据往往与实验条件和状态关系密切,难于掌控,并且实验量非常大,获得实验数据有限。因此,需要探索通过建立理论计算模型,以实验数据作为检验来计算分析结合面法向接触刚度及其与影响因素间的关系。目前,在结合面法向接触问题理论计算模型研究方面主要有两类方法:(1)基于分形几何理论的模型。Majumder 等[6,7]提出了接触分形理论和接触分形模型。Greenwood 等[8,9]基于赫兹理论,第一个建立粗糙表面的弹性接触模型(简称 GW 模型),但 GW 模型局限于处理低塑形指数的表面接触,假设的微凸体的变形机制为完全弹性。张学良、黄玉美和傅卫平[3,10,11]在一定的假设下,基于接触分形理论,直观地揭示了结合面法向接触刚度与结合面诸参数之间的非线性关系,但只是局限于法向接触刚度与法向载荷的仿真研究,未与实验数据进行定量比较。Jiang 等人[10-14]利用接触分形理论,建立结合面法向接触刚度的分形模型,并通过实验方法将实验数据和模型仿真数据进行对比研究。但是接触面积、接触载荷等参数随着表面微凸体的变形而发生变化,导致用分形理论计算结合面法向刚度与实际结果之间误差较大。(2)基于接触力学理论的模型。Zhao 等人[15-18]基于接触力学理论,提出一种新的粗糙表面弹-塑性微观接触模型(简称 ZMC 模型),接触理论建立在对粗糙表面微观形貌特征的传统定量化统计描述结果的基础之上,但仅局限于对粗糙表面微凸体弹性、弹塑性、塑性接触变形、接触面积和接触载荷等方面的研究[19-21],并且没有研究结合面法向接触刚度及其与法向载荷之间的关系。

综合考虑上述两种模型,本文利用分形几何理论获得结合面表面形貌特征参数,基于 ZMC 接触模型,引入塑性指数对接触参数的影响规律,建立机械粗糙表面弹塑性结合面的法向接触刚度模型;分析一定的塑性指数下结合面法向接触刚度间与法向载荷的关系;以文献[12]中的实验数据作为算例,对本文的理论模型的计算结果进行实验验证,并与其理论模型(简称 JZZ 模型)计算结果进行分析比较。

1 法向接触刚度计算模型

机械加工表面在微观尺度上存在大量的微凸体,两个机械加工粗糙表面接触仅发生在微凸体上,因此,研究表面微凸体的接触特性是结合面接触的基础。以下首先对接触表面进行假设,在此基础上,应用分形几何理论建立表面微凸体模型;然后,运用接触力学理论,描述单个微凸体从弹性、弹塑性到塑形变形的规律;最后,应用微凸体统计理论建立法向接触载荷和接触刚度模型。

1.1 模型假设

两个粗糙表面的接触可以简化成一个刚性光滑平面与一个等同粗糙表面接触,如图 1 所示,模型假设与 GW 模型和 CEB 模型所做的假设相同[8,15]。其中 y_s 为表面平均高度平面和微凸体平均高度平面之间的距离,R 为微凸体的曲率半径。当在刚性平面施加载荷时,微凸体将发生变形,其法向变形量为 ω,z 和 d 分别是此微凸体的高度和两表面间的距离。h 是刚性平面与表面平均高度平面间的距离。微凸体的高度服从高斯分布,法向变形量表示为:

$$\omega = z - d \tag{1}$$

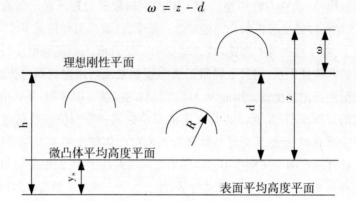

图 1　结合面接触示意图

1.2 表面微凸体模型

根据分形理论,机械加工表面轮廓具有非平稳性、自相似性和多重尺度特性,其表面微凸体顶端等效曲率半径 R 和高度方差 σ 可由下式计算:

$$\sigma = \left(\frac{1}{(2\ln\gamma)(4-2D)}\right)^{1/2} G^{(D-1)} L^{2-D} \tag{2}$$

$$a' = \left[2^{9-2D} \pi^{D-3} \lambda^{-2} G^{2D-2} \left(\frac{E}{H}\right)^2 \ln\gamma\right]^{1/(D-1)} \tag{3}$$

$$R = \frac{(a')^{D/2}}{2^{4-2D} \pi^{D/2} G^{D-1} (\ln\gamma)^{0.5}} \tag{4}$$

式中 G ——反映轮廓大小的特征尺度系数,γ ——表示随机轮廓的空间频率,$\gamma = 1.5$

D ——轮廓的分形维数,L ——采样长度,H ——材料的硬度,$H = 2.8Y$,Y ——材料的屈服强度值

1.3 单个微凸体接触刚度

根据接触力学理论,微凸体表面刚度、接触载荷等参数与其变形量 ω 有密切关系,变形过程从完全弹性到弹 - 塑性最终转化为完全塑性的变形。

(1)完全弹性接触

根据赫兹接触理论,微凸体变形量较小时,发生弹性变形,接触载荷 w_e 和接触刚度 k_e 表示为[18]:

$$w_e = \frac{4}{3}ER^{1/2}\omega^{3/2} \tag{5}$$

$$k_e = 2ER^{1/2}\omega^{1/2} \tag{6}$$

式中:E 为等效弹性模量,可以用两接触表面的弹性模量值 E_1、E_2 和泊松比 ν_1、ν_2 表示:

$$\frac{1}{E} = \frac{1-\nu_1^2}{E_1} + \frac{1-\nu_2^2}{E_2} \tag{7}$$

随着微凸体变形量 ω 增加,将进入弹-塑性变形阶段,从完全弹性变形向弹-塑性变形转变的临界变形值 ω_e 与粗糙表面的材料属性有关[21],

$$\omega_e = \left(\frac{\pi CY}{2E}\right)^2 R \tag{8}$$

式中 $C = 1.295\exp(0.736\nu)$

(2)完全塑性接触

当 $\omega \geq \omega_p$ 时,微凸体将发生完全塑性变形,ω_p 为弹-塑性接触变形和完全塑性接触变形的临界值,由文献[21]可得,$\omega_p = 110\omega_e$。接触载荷 w_p 和接触刚度 k_p 表示为:

$$w_p = 2\pi RH\omega \tag{9}$$

$$k_p = 2\pi RH \tag{10}$$

(3)弹-塑性接触

当 $\omega_e \leq \omega < \omega_p$ 时,微凸体变形是弹性和塑性共存变形阶段,依据文献[16]的接触理论模型,接触载荷 w_{ep} 和接触刚度 k_{ep} 表示为:

$$w_{ep} = \pi HR\omega\left[1 - 1(1-\lambda)\frac{\ln\omega_p - \ln\omega}{\ln\omega_p - \ln\omega_e}\right]f_1(\omega) \tag{11}$$

$$k_{ep} = \pi RH\omega\left[1 - (1-\lambda)\frac{\ln\omega_p - \ln\omega}{\ln\omega_p - \ln\omega_e}\right]f_2(\omega) + \pi RH\left[(1-\lambda)\frac{1}{\ln\omega_p - \ln\omega_e}\right]f_1(\omega)$$

$$+ \pi RH\left[1 - (1-\lambda)\frac{\ln\omega_p - \ln\omega}{\ln\omega_p - \ln\omega_e}\right]f_1(\omega) \tag{12}$$

其中:$f_1(\omega) = 1 - 2\left(\frac{\omega-\omega_e}{\omega_p-\omega_e}\right)^3 + 3\left(\frac{\omega-\omega_e}{\omega_p-\omega_e}\right)^2$,$f_2(\omega) = 6\frac{\omega-\omega_e}{\omega_p-\omega_e} - 6\left(\frac{\omega-\omega_e}{\omega_p-\omega_e}\right)^2$

1.4 两个粗糙结合表面法向接触刚度

两个宏观粗糙表面接触微观上表现为微凸体的接触,如果在名义接触面积 A 上有 N 个微凸体,根据以上微凸体的三种变形机制,两表面总的接触载荷 W_n 和接触刚度 K_n 分别是所有接触微凸体接触载荷和接触刚度之和。同时,大量研究表明,工程表面的微凸体高度服从高斯分布,对于给定某一表面距离 d,接触载荷 W_n 和接触刚度 K_n 可表示为:

$$W_n(d) = w_e(d) + w_{ep}(d) + w_p(d)$$

$$= N\int_{d}^{d+\omega_e} w_e \Phi(z)dz + N\int_{d+\omega_e}^{d+\omega_p} w_{ep}\Phi(z)dz + N\int_{d+\omega_p}^{\infty} w_p\Phi(z)dz$$

$$= \frac{4}{3}\eta AER^{1/2}\int_{d}^{d+\omega_e}\omega^{3/2}\Phi(z)dz + \eta A\pi RH\int_{d+\omega_e}^{d+\omega_p}f_1(\omega) \times$$

$$\left[1 - (1-\lambda)\frac{\ln\omega_p - \ln\omega}{\ln\omega_p - \ln\omega_e}\right]\omega\Phi(z)dz + 2\pi\eta AHR\int_{d+\omega_p}^{\infty}\omega\Phi(z)dz \quad (13)$$

$$K_n(d) = k_e(d) + k_{ep}(d) + k_p(d)$$

$$= N\int_{d}^{d+\omega_e}k_e\Phi(z)dz + N\int_{d+\omega_e}^{d+\omega_p}k_{ep}\Phi(z)dz + N\int_{d+\omega_p}^{\infty}k_p\Phi(z)dz$$

$$= 2\eta AER^{1/2}\int_{d}^{d+\omega_e}\omega^{1/2}\Phi(z)dz + 2\pi\eta ARH\int_{d+w_p}^{\infty}\Phi(z)dz$$

$$+ \eta A\pi RH\int_{d+\omega_e}^{d+\omega_p}\left[1-(1-\lambda)\frac{\ln\omega_p-\ln\omega}{\ln\omega_p-\ln\omega_e}\right]f_2(\omega)\omega\Phi(z)dz$$

$$+ \eta A\pi RH\int_{d+\omega_e}^{d+\omega_p}f_1(\omega)\left[(1-\lambda)\frac{1}{\ln\omega_p-\ln\omega_e}\right]\Phi(z)dz$$

$$+ \eta A\pi RH\int_{d+\omega_e}^{d+\omega_p}\left[1-(1-\lambda)\frac{\ln\omega_p-\ln\omega}{\ln\omega_p-\ln\omega_e}\right]f_1(\omega)\Phi(z)dz \quad (14)$$

式中　η ——微凸体分布密度

　　　$\Phi(z)$ ——微凸体高度分布概率密度函数

　　　λ ——平均接触压力系数,$\lambda = 0.4645 + 0.3141\nu$

为了使模型的计算结果有广泛的通用性,对模型中的参数进行无量纲化。无量纲化后的接触载荷和接触刚度分别为:

$$W_n^* = \frac{4}{3}\beta\left(\frac{\sigma}{R}\right)^{0.5}\int_{h^*-y_s^*}^{h^*-y_s^*+\omega_e^*}\omega^{*3/2}\Phi^*(z^*)dz^* + \frac{2\pi H\beta}{E}\int_{h^*-y_s^*+\omega_p^*}^{\infty}\omega^*\Phi^*(z^*)dz^*$$

$$+ \frac{\pi H\beta}{E}\int_{h^*-y_s^*+\omega_e^*}^{h^*-y_s^*+\omega_p^*}\omega^*f_1^*(\omega^*)\left[1-(1-\lambda)\frac{\ln\omega_p^*-\ln\omega^*}{\ln\omega_p^*-\ln\omega_e^*}\right]\Phi^*(z^*)dz^* \quad (15)$$

$$K_n^* = \frac{2\beta\sqrt{A}}{\sqrt{\sigma R}}\int_{h^*-y_s^*}^{h^*-y_s^*+\omega_e^*}\omega^{*1/2}\Phi^*(z^*)dz^* + \frac{2\pi\beta H\sqrt{A}}{E\sigma}\int_{h^*-y_s^*+\omega_p^*}^{\infty}\Phi(z)dz$$

$$+ \frac{\pi\beta H \sqrt{A}}{E} \int_{h^* - y_s^* + \omega_e^*}^{h^* - y_s^* + \omega_p^*} \left[1 - (1-\lambda) \frac{\ln\omega_p^* - \ln\omega^*}{\ln\omega_p^* - \ln\omega_e^*} \right] f_2^*(\omega^*) \omega^* \Phi^*(z^*) dz^*$$

$$+ \frac{\pi\beta H \sqrt{A}}{E\sigma} \int_{h^* - y_s^* + \omega_e^*}^{h^* - y_s^* + \omega_p^*} f_1^*(\omega^*) \left[(1-\lambda) \frac{1}{\ln\omega_p^* - \ln\omega_e^*} \right] \Phi^*(z^*) dz^*$$

$$+ \frac{\pi\beta H \sqrt{A}}{E\sigma} \int_{h^* - y_s^* + \omega_e^*}^{h^* - y_s^* + \omega_p^*} \left[1 - (1-\lambda) \frac{\ln\omega_p^* - \ln\omega^*}{\ln\omega_p^* - \ln\omega_e^*} \right] f_1^*(\omega^*) \Phi^*(z^*) dz^* \quad (16)$$

其中:$f_1^*(\omega^*) = 1 - 2\left(\frac{\omega^* - \omega_e^*}{\omega_p^* - \omega_e^*}\right)^3 + 3\left(\frac{\omega^* - \omega_e^*}{\omega_p^* - \omega_e^*}\right)^3$, $f_2^*(\omega^*) = 6\frac{\omega^* - \omega_e^*}{\omega_p^* - \omega_e^*}$

$- 6\left(\frac{\omega^* - \omega_e^*}{\omega_p^* - \omega_e^*}\right)^2$, $\beta = \eta\sigma R$, $\omega^* = z^* - h^* + y_s^*$

$y_s^* = 0.045944/\beta$, $\Phi^*(z^*) = 1/\sqrt{2\pi}(\sigma/R)\exp[-0.5(\sigma/\sigma_s)^2 z^{*2}]$

式中 h^* —无量纲平均分离距离 $h^* = h/\sigma$,y_s^* —无量纲微凸体平均高度线与表面高度平均线距离 $y_s^* = y_s/\sigma$

ω^* —无量纲法向变形量 $\omega^* = \omega/\sigma$,$\Phi^*(z^*)$ —无量纲表面微凸体高度分布函数

z^* —无量纲表面微凸体高度 $z^* = z/\sigma$,ω_e^*,ω_p^* —分别为无量纲完全弹性和完全塑性法向临界变形量 $\omega_e^* = \omega_e/\sigma$,$\omega_p^* = \omega_p/\sigma$

1.5 塑性指数

由以上理论分析可知,表面微凸体接触变形随着法向载荷的增加经历了三个变形过程,如何衡量其变形触状态是解决该问题的关键。在 G-W 模型中提出了塑性指数的概念,塑性指数 ψ 作为衡量弹性还是塑性接触状态的判据,可以通过 ψ 将材料本身的物理性能与接触面的几何形状联系起来。

表面粗糙程度由参数 β 和 σ/R 值确定,其值是通过典型的工程表面由实验的方法获得。材料性能参数为:$E_1 = E_2 = 207 \text{GPa}$,$H = 1.96 \text{GPa}$,泊松比 $\nu = \nu_1 = \nu_2 = 0.29$,塑性指数 ψ 与参数 β、σ/R 之间的对应关系如表1所示[16]。

$$\psi = \frac{2E}{1.5\pi\lambda H}\left(\frac{\sigma}{R}\right)^{0.5}\left(\frac{3.717 \times 10^{-4}}{\beta^2}\right)^{0.25} \quad (17)$$

表1 表面工程参数和塑性指数

序号	σ/R	β	ψ
1	8.75×10^{-5}	0.0302	0.5
2	1.6×10^{-4}	0.0339	0.7

续表

序号	σ/R	β	ψ
3	3.02×10^{-4}	0.0414	1.0
4	1.144×10^{-3}	0.0541	2.0

2　理论结果与分析

塑性指数的大小反映了微凸体的塑性变形尺度,其大小依赖接触表面的力学特性和表面微观形貌特性。以下分析了两个粗糙表面在表 1 中不同塑性指数下的接触刚度与法向载荷间的特性关系。

图 2 给出了在不同塑性指数下无量纲法向接触载荷与无量纲法向接触刚度之间的关系曲线。对比四个塑性指数下的曲线可以看出,法向接触刚度随法向载荷的增大而单调增大。当 $\psi = 0.5$ 情况下,法向载荷与法刚度之间近似为线性关系,随着 ψ 值的增大,法向载荷与法刚度之间的关系表现出较强的非线性关系,这是合乎道理的,因为随着 ψ 的增大,在同样的载荷下有更多的接触微凸体进入塑性屈服阶段,导致接触刚度相对降低。法向载荷与法向刚度之间服从 $K_n = aP^b$(其中 a,b 为常数)曲线关系,这与文献[2、3]中实验结论相一致。

从以上不同塑性指数下的曲线对比表明,当塑性指数较小时,接触微凸体主要以弹性变形为主,法向载荷与法刚度之间表现为较弱的非线性关系,随着塑性指数的增加,非线性的关系表现得比较明显,此时,微凸体的变形从弹性变形为主向塑性变形为主转变。

3　实验结果对比与分析

为了说明本文模型的有效性,本节作进一步实验验证。

文献[12]建立了基于分形接触理论的结合面法向接触刚度模型,并通过实验验证了 JZZ 模型的有效性。实验参数如下:材料为铸铁,材料弹性模量 $E = 100\text{GPa}$,泊松比 $\nu_1 = \nu_2 = 0.25$,硬度 $H = 2.07\text{GPa}$。接触表面分别采用铣削加工和磨削加工,铣削加工表面的分形维数 D 和高度尺度参数 G 分别为:1.2183 和 1.2117×10^{-14},磨削加工表面的分形维数 D 和高度尺度参数 G 分别为:1.4058 和 9.7582×10^{-11}。

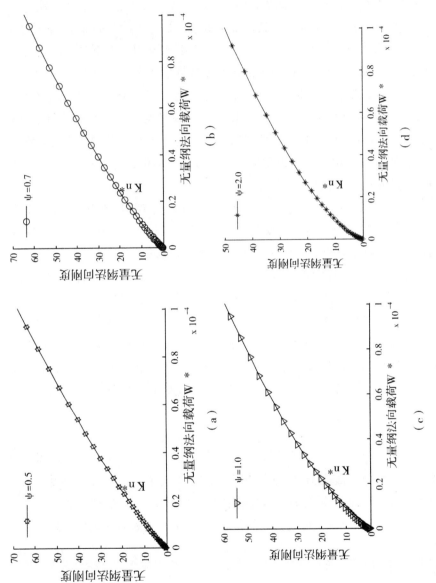

图 2 不同塑性指数下法向接触载荷与接触刚度之间的关系

本文应用文献[12]中的实验数据,进一步验证该模型所建立的法向载荷和法向刚度之间关系的有效性,具体方法如下:将上述实验参数分别代入以上相应式子中,可得到铣削加工 $\sigma/R = 0.0002$,$\psi = 3.22$;磨削加工 $\sigma/R = 0.0014$,$\psi = 4.07$。由式(9)、(10)分别建立法向接触面压和法向接触刚度之间的关系曲线。

图3(a)、(b)分别是不同表面铣削加工和磨削加工下的实验曲线、本模型曲线和JZZ模型曲线对比关系图,由图3(a)可以看出,随着法向接触压力的增大,本模型曲线和实验曲线走势一致,法向接触刚度计算值和实验值接近,但是JZZ模型曲线与实验曲线误差越来越大。当法向压力值小于1.8MPa时,本模型的计算值大于实验值,当法向压力值大于1.8MPa时,本模型的计算值与实验值基本相等,很好地计算了法向接触刚度。图3(b)同样表现出与3(a)相同的曲线走势,当法向压力值小于1.2MPa时,本模型的计算值大于实验值,当法向压力值大于1.2MPa时,本模型的计算值小于实验值。

对比结算结果和实验曲线,可以得出以下结论:(1)铣削、磨削加工表面塑性指数分别为 $\psi = 3.22$,$\psi = 4.07$,塑性指数的计算结果与文献[21]"结合面材料越软,表面粗糙度越大,塑性指数越高"的结论一致,表面该模型计算方法的可行性。(2)结合面法向接触刚度随着无量纲法向接触面压的增大而增大,即法向载荷增大有利于提高结合面的法向接触刚度;(3)在同样载荷下,铣削加工表面的法向接触刚度比磨削加工表面接触刚度小,这是因为铣削加工的表面粗糙度值比磨削加工的表面粗糙度值大,分别是 $Ra = 2.47~\mu m$ 和 $Ra = 1.02~\mu m$,当两个表面接触时,磨削加工表面微凸体接触的数量比铣削加工微凸体接触数量多,而且微凸体的高度也有所差异,导致接触刚度增大。(4)本模型与实验曲线吻合程度相对较好。

4 结论

(1)将分形几何理论和接触力学理论相结合,提出了结合面法向接触刚度计算模型,该模型更全面地描述了粗糙表面微凸体变形过程中法向接触刚度的数值计算方法,为进一步研究结合面动态接触刚度模型提供了一定的理论基础。

(2)塑性指数是决定粗糙表面间的接触性能,也是判断粗糙表面接触时产生弹塑性变形的主要参数,它将表面的物理机械性能和微凸体几何形状结合起来,反映配合表面接触的力学性质。

(3)分析结果表明,结合面法向接触刚度随着法向接触载荷的增加而单调增大。在一定的条件下,随着塑性指数的增加,结合面法向接触刚度与法向接触载荷间的关系由线性关系变为非线性关系,微凸体由弹性变形为主向弹塑性变形为主最后转变为以塑性变形为主。

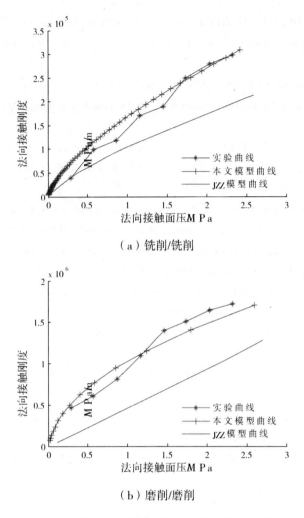

图3 不同加工下法向接触面压和法向接触刚度之间的关系

(4)数值仿真结果与实验结果之间的一致性说明了本文所建的法向接触刚度模型是合理的。

参考文献

[1]REN Y,BEARDS C F. Identification of effective linear joints using coupling and joint identification techniques [J]. ASME Journal of Vibration and Acoustics,1998,120(2):331-338.

[2]黄玉美,傅卫平,佟浚贤.结合面法向动态特性参数研究[J].机械工程学报,1993,29(3):74-78.

HUANG Y M,FU W P,TONG J X. Research on the normal dynamic characteristics parameters of machine joint surfaces[J]. Chinese journal of mechanical engineering,1993,29(3):74-78.

[3] FU W P, HUANG Y M, ZHANG X L et al. Experimental investigation of dynamic normal characteristics of machined joint surfaces[J]. ASME Journal of Vibration and Acoustics, 2000, 122(4): 393-398.

[4] SHI X, ANDREAS A, POLYCARPOU. Investigation of contact stiffness and contact damping for magnetic storage head – disk interfaces[J]. ASME Journal of Tribology, 2008, 130/021901: 1-9.

[5] KONRAD, KONOWALSKI. Experimental research and molding of normal contact stiffness and contact damping of machined joint surfaces[J]. Advances in Manufacturing Science and Technology, 2009, 3(33): 53-68.

[6] MAJUMDAR A, BHUSHAN B. Role of fractal geometry in roughness characterization and contact mechanics of surfaces[J]. ASME Journal of Tribology, 1990, 112(1), 205-216.

[7] MAJUMDAR A, BHUSHAN B. Fractal model of elastic – plastic contact between rough surfaces[J]. ASME, Journal of Tribology 1991, 113(1): 1-11.

[8] GREENWOOD J A, WILLIAMSON J P B. Contact of nominally flat surfaces[J]. Proceedings of the Royal Society, 1966. A952; 300-319.

[9] CIAVARELLA M, GREENWOOD J. Inclusion of "interaction" in theGreenwood and Williamson contact theory [J]. Wear, 2008, 265(5): 729-734.

[10] 张学良,黄玉美,傅卫平. 粗糙表面法向接触刚度的分形模型[J]. 应用力学学报, 2000, 17(2): 31-35.
ZHANG X L, HUANG Y M, FU W P. Fractal Model of Normal Contact Stiffness between Rough Surfaces[J]. Chinese Journal of Applied Mechanics, 2000, 17(2): 31-35.

[11] 尤晋闽,陈天宁. 结合面法向动态参数的分形模型[J]. 西安交通大学学报, 2009, 9(43): 91-94.
YOU J M, CHEN T N. Fractal model for normal dynamic parameters of joint surfaces[J]. Journal of Xi'an Jiaotong University, 2009, 9(43): 91-94.

[12] JIANG S Y, ZHENG Y J, ZHU H. A contact stiffness model of machined plane joint based on fractal theory[J]. Journal of Tribology, 2010, 132/011401: 1-7.

[13] 杨红平,傅卫平等. 小波系数表征机械加工表面分形特征的计算方法[J]. 仪器仪表学报, 2010, 31(7): 1454-1459.
YANG H P, FU W P, et al. Calculation method for fractal characteristics of machining topography surface based on wavelet coefficients[J]. Chinese Journal of Scientific Instrument, 2010, 31(7): 1454-1459.

[14] KOMVOPOULOS K, YE N. Three – dimensional contact analysis of elastic – Plastic layered media with fractal surface topographies[J]. Journal of Tribology, 2001, 123(3): 632-640.

[15] 赵永武,吕彦明,蒋建忠. 新的粗糙表面弹塑性接触力学模型[J]. 机械工程学报, 2007, 43(3): 95-101.

ZHAO Y W, LU Y M, JIANG J ZH. New elastic – plastic model for the contact of rough surfaces[J]. Chinese journal of mechanical engineering, 2007, 43(3):95 – 101.

[16] ZHAO Y W, DAVID D M, CHANG L. An asperity microcontact model incorporating the transition from elastic deformation to fully plastic flow[J]. ASME Journal of Tribology, 2000, 122:86 – 93.

[17] JENG YEAU – REN, PENG SHIN – RUNG. Elastic – plastic contact behavior considering asperity interactions for surfaces with various height distributions[J]. ASME Journal of Tribology, 2006, 126:620 – 625.

[18] ALMQVIST A, SAHLIN F, LARSSON R, et al. On the dry elastic – plastic contact of nominally flat surfaces[J]. ASME Journal of Tribology, 2007, 40(10):574 – 579.

[19] KOGUT L, ETSION I. A static friction model for elastic – plastic contacting rough surface[J]. ASME Journal of Tribology, 2004, 126(1):34 – 40.

[20] ROBERT L JACKSON, JEFFREY L. STREATOR. A multi – scale model for contact between rough surfaces[J]. Wear, 2006, 261:1337 – 1347.

[21] ROBERT L. JACKSON, ITZHAK GREEN. A statistical model of elasto – plastic asperity contact between[J]. Journal of Tribology International Transactions, 2006, 46(3):906 – 914.

注：本文曾发表在2013年《机械工程学报》第1期上

基于回声状态网络的结合面特性参数建模

杨红平*

机械结构中存在大量结合面,在机床静态变形中,由各结合面引起的变形量高达85%,各种结合面的接触刚度约占机床总刚度的60-80%。本文针对机械结合面接触特性参数,提出基于ESN回声状态网络理论对机械结合面法向接触刚度进行仿生学建模。以4种组合条件下的结合面接触刚度为算例,采用算法学习训练域和预测域相分离的方法,在对影响法向接触刚度的主要因素的定量化处理的基础上,进行ESN算法建模和计算结果误差分析,结果表明,该算法的预测精度可达0.0016%以上。同时,在同等条件下,通过该算法与BP神经网络、RBP神经网络、MPSO-BP网络算法预测能力比较分析,结果表明,回声状态网络计算精度最高,并将该建模计算结果进行工程应用。

1 引言

在机床及机械结构中存在着大量的结合面,研究表明[1-2],机床中结合面的刚度约占机床总体刚度的60-80%。因此,研究机械结合面的接触刚度,为机床及其它机械结构的优化设计提供必要的理论基础。

影响机械结合面接触刚度的因素主要有材料属性、载荷性质、加工工艺、表面粗糙度、介质等,并且存在复杂的非线性映射关系。研究机械结合面接触刚度的方法主要有:(1)理论计算。Jeng[3]基于GW(Greenwood – Williamson)[4]接触模型,通过对机械加工粗糙表面形貌进行统计描述,建立粗糙表面弹塑性微观接触模型。Majumdar[5]采用分形函数表征粗糙表面形貌,首次提出基于分形几何理论的结合面接触刚度模型。之后,尤晋闽等[6-7]在综合考虑粗糙表面微凸体在完全弹性、弹塑性和完全塑性三种变形机理的理论计算基础上,对结合面法向刚度进

---

* 作者简介:杨红平(1974—),男,甘肃秦安人,天水师范学院教授,博士(后),主要从事机电系统非线性动力学研究。

行数值仿真分析。Song和Etsion[8]在分析真实粗糙表面微凸体三维接触特性基础上,通过有限元理论,建立了结合面接触刚度与接触面积、接触载荷的接触参数的表达式。然而,影响结合面特性参数因素太多,受到公式使用条件的限制,只能在满足特定条件的情况下才能使用。(2)实验获取。Fu[9]对机械结合面法向刚度和阻尼特性进行大量实验研究,分析结合面材料、加工方法、介质等对法向接触刚度的影响。Pitenis[10]对结合面基础特性参数进行了大量的研究,通过实验与理论结合的方法得到结合面的刚度特性公式,并对典型的机械结构进行优化。该方法仅适于少影响因素问题的描述,而对多影响因素问题,有时难以处理、拟合。(3)仿生学预测。Sonbaty[11-12]等在分析了影响结合面基础特性参数的众多因素,通过对训练完成后的BP(back propagation)网络权值,建立了结合面静态特性参数与其连续变化影响因素之间的非线性关系。杨红平[13]等提出用改进粒子群优化算法优化BP神经网络参数,实现了结合面接触刚度的预测。朱坚民[14]提出一种改进的自适应遗传算法用于固定结合面动态特性参数的识别,获得了较高的识别精度。然而,BP神经网络算法使用梯度下降学习算法,神经网络的规模不能太大,容易使得算法出现局部极小化。粒子群算法对离散的优化问题处理不佳,容易陷入局部最优。而回声状态网络算法(Echo State Network,ESN)抛弃了BP神经网络梯度下降算法,分离了递归神经网络的信息表达与权值训练过程,在拥有很强的非线性系统逼近能力的同时,极大地简化了递归神经网络的训练,实现了对样本数据非线性时间序列预测[15-16]。

  本文提出将回声状态网络算法应用到机械结合面接触特性参数的建模与预测中,在分析算法原理的基础上,通过对影响结合面特性参数的主要因素和量化的解析,对一定实验条件下的结合面法向接触刚度进行建模预测与误差分析,并得到实际工程应用。

2 回声状态网络
2.1 回声状态网络的基本原理

  回声状态网络算法是2004年由德国学者Jaeger[17]在《Science》提出的新型递归神经网络,其主要突出的特点体现在以下几个方面:(1)在学习训练的方法上抛弃了BP神经网络梯度下降的算法,采用动态递归的神经网络算法。(2)网络结构由输入层、中间层、输出层组成,输入层由K个输入神经元组成,中间层由N个动态池(Dynamic Reservoi,DR)神经元组成,输出层由L个输出神经元组成,如图1所示。

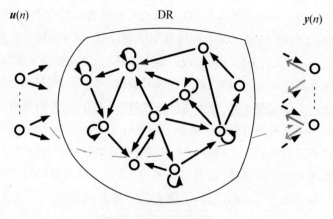

图1　ESN基本结构

中间层由数量众多的动态池(DR)神经元相互连接构成,动态池作为回声状态网络最重要的部分,将神经元随机连接形成复杂的神经元网络,对外界环境的刺激进行学习并做出相应的反应。图1中实线连接在网络初始化时产生,无需改变,虚线部分需要在训练中学习。假设网络在 n 时刻的输入层、中间层和输出层的序列向量分别表示为:

$$u(n) = (u_1(n),\cdots,u_K(n)) \quad (1)$$

$$x(n) = (x_1(n),\cdots,x_N(n)) \quad (2)$$

$$y(n) = (y_1(n),\cdots,y_L(n)) \quad (3)$$

ESN内更新方程和输出方程表达式分别为:

$$x(n+1) = f(W^{in}u(n+1) + Wx(n) + W^{back}y(n)) \quad (4)$$

$$y(n+1) = f^{out}(W^{out}(u(n+1),x(n+1),y(n))) \quad (5)$$

其中, $f = [f_1, f_2, \cdots, f_N]^T$, $f^{out} = [f_1^{out}, f_2^{out}, \cdots, f_L^{out}]^T$ 分别为动态池内部神经元的激活函数和输出层神经元激活函数,一般地, f 取双曲正切函数 $\tanh(\cdot)$, f^{out} 取对称型sigmoid函数; $W_{N\times K}^{in}$ 是输入层到中间层的神经元权矩阵; $W_{N\times N}$ 是中间层内部权矩阵; $W_{N\times L}^{back}$ 是输出层到中间层的神经元权矩阵; $W_{L\times(K+N)}^{out}$ 是中间层神经元到输出层权矩阵。

2.2　ESN的学习算法实现步骤

ESN的建模与预测过程遵循机器学习的规律,通过对样本数据的观察与学习,进一步完成样本数据规律的深层次挖掘,完成目标任务的建模与预测。

ESN神经网络训练的基本思想是[17-18]:给定长度为 T 的输入、输出训练样本序列,通过 W^{in} 和 W^{back} 激励,DR中的内部神经元,使DR产生回声震荡,以实现对信息的短期记忆,使网络输出 $y(n)$ 逼近期望输出 $d(n)$,最后通过线性回归使训

练均方误差最小化从而得到 W^{out}。ESN 网络训练算法如下:

(1)根据数据集规模确定输入层、中间层、输出层的向量阵 K、N、L。

(2)ESN 网络初始化,其中使 W 稀疏且谱半径 $\rho < 1$,初始时刻取 $x(0) = 0$,$d(0) = 0$,通过训练样本由式(4)驱动网络。

(3)随机产生权矩阵 W^{in}、W、W^{back},对于长度为 T 的输入、输出训练样本,W^{in}、W^{back} 使 DR 产生回声震荡,对信息进行短期记忆,使得网络输出 $y(n)$ 逼近期望输出 $d(n)$,最终得到 W^{out}。为了保证 DR 的回声状态特性,网络从任意初始化状态 $x(0)$ 开始,最终收敛于0,权矩阵 W 的谱半径小于1。

(4)定义矩阵 $M_{T \times (K+N+L)}$,$T_{T \times L}$,其中 M 由每一时刻各输出的状态阵 $(u(n)$,$x(n),y(n))$,记录了输入神经元,动态池神经元和输出神经元的历史状态矩阵,T 的行向量由期望输出 $d(n)$ 在每一时刻的反双曲正切函数 $\operatorname{arctanh} d(n)$ 组成。

(5)输出权值的计算,表达式为:

$$W^{out} = (M^{-1}T)^{T} \tag{6}$$

网络算法的序列输出由式(5)获得。

(6)ESN 训练结束后,采用标准均方误差网络测试,判断均方差是否达到指定的允许误差,其表达式为:

$$\sigma_{MSE} = \frac{1}{T}\sum_{n=1}^{T}(\operatorname{arctanh}(d(n)) - \operatorname{arctanh}(y(n)))^{2} \tag{7}$$

3 机械结合面接触特性参数建模与分析

3.1 ESN 输入和输出参数分析与量化处理

机械结合面的接触刚度受诸多因素的影响,如结合面接触面压、接触表面加工方法、配对副材料、接触表面粗糙度、结合面间介质等,这些因素之间存在着复杂的非线性映射关系。根据影响结合面特性参数的主要因素,以结合面接触面压、接触表面加工方法、配对副材料、接触表面粗糙度、结合面间介质等 5 个影响因素为输入参数,法向接触刚度为输出参数。

机械结合面接触刚度的影响因素的量纲、类型方面存在较大差异,接触面压、接触刚度、粗糙度采用本身参数量化,而结合面材料、介质和加工方法没有量化参数,需要根据其特性进行数据定量化处理。对于材料而言,影响其力学性能的主要参数是弹性模量,涉及的主要材料有灰铸铁、45 钢、塑料等,对于配对副材料一般采用复合弹性模量来描述,其表达式为:

$$\frac{1}{E} = \frac{1 - \nu_{1}^{2}}{E_{1}} + \frac{1 - \nu_{2}^{2}}{E_{2}} \tag{8}$$

其中，E 为等效弹性模量可以用两接触表面的弹性模量值 E_1、E_2 和泊松比 ν_1、ν_2。

对于加工方法而言，机械结合表面的加工方法主要有车削、铣削、刮削和磨削，不同的加工工艺表现出的粗糙表面完整度。研究表明[5-7]，机械表面的轮廓分形维数在[1,2]之间，加工精度越高，表面微观结构越精细，分形维数越大，因此，采用分形维数来定量描述加工方法。对于结合面介质而言，只有无油和有油两种情况，本文将介质定量为无油为0，有油为1。

表1、表2分别为配对副材料、表面加工方法的定量化描述。

表 1　配对副材料定量化描述

表面加工方法	定量化描述
铸铁－贴塑	60
铸铁－铸铁	90
45 钢－铸铁	110

表 2　配对副表面加工方法定量化描述

配对副材料	定量化描述
磨削－磨削	1.8
铣削－刮削	1.6

3.2　实验组合条件描述

表3列出了部分结合面特性参数的实验组合条件，接触刚度类型为法向，各类型共有四组，配对副材料主要是在机床行业中常见结合面工作材料，主要有：铸铁－铸铁、45钢－铸铁、铸铁－贴塑，加工方法方面主要有磨削－磨削、铣削－刮削，在结合面介质上列举了20油摩擦和无油干摩擦两种。

表 3　结合面法向接触实验组合条件

序号	配对副材料	表面粗糙度	加工方法	介质
1	铸铁－贴塑	0.65－2.4	磨削－磨削	无油
2	45 钢－铸铁	0.3－0.4	磨削－磨削	无油
3	铸铁－铸铁	0.4－0.4	磨削－磨削	无油
4	铸铁－铸铁	0.4－0.4	铣削－刮削	20 号油

3.3 ESN 训练样本参数设计

模型的输入量是影响结合面法向接触刚度的定量化参数,即结合面法向面压、表面粗糙度、材料特性即材料弹性模量、结合面加工方法和结合面间介质,输出量为结合面法向接触刚度。

首先给定 ESN 储备池的参数,即 DR 神经元个数 N、DR 稀疏度 sd、DR 中 W 的谱半径 ρ 以及 DR 输入权值放缩尺度 IS,依照经验取值和多次实验得到参数最优组合,参数设置如表 4 所示。

表 4 ESN 储备池参数设置

DR 参数	取值
DR 维数 N	100
DR 稀疏度 sd	2%
DR 中 w 的谱半径 ρ	0.8
DR 输入权值放缩尺度 IS	0.05

开始训练网络,此时 $n=0$,初始状态 $x(0)=0$、$d(0)=0$,ESN 内部神经元状态按照(1)式更新。当 $n=1$ 时网络运行并通过式(4)获取 ESN 训练输出权矩阵 W^{out};网络最终输出由式(5)得到。将 tanh 函数作为内部单元输出函数和输出函数,利用训练得到的输出权矩阵 W^{out} 测试网络。

4 结合面法向接触刚度仿真实验及结果分析

现有文献[12-14]对结合面特性参数的建模与预测思想:在一定的实验参数范围内进行训练学习和预测,再将实验曲线和预测曲线进行对比分析,这种建模预测的不足在于:预测值只能反映训练域的情况,对训练域以外的预测可能会反映实验值,有可能偏离实验值,这样不能有效达到预测效果。

本文基于 ESN 回声状态网络的结合面法向接触刚度的建模与预测思想是:选取实验域内的一定实验参数进行回声状态网络的训练与学习,建立一定条件下的结合面特性参数与接触面压之间的非线性映射关系,然后根据训练学习结果预测训练域以外的结合面特性参数与接触面压之间的关系,并对预测结果与实验结果进行误差和方差评价分析。这种做法在一定程度上能有效评价该算法的预测能力和可行性。

4.1 法向接触刚度预测结果与误差分析

图 2 是在表 1 的组合条件下的结合面法向接触刚度的回声网络算法的训练

与预测曲线。由图2(a)中结合面法向载荷和法向刚度的预测曲线和实验曲线对比,训练法向载荷的取值范围是 0－0.9MPa,从图中可以看出,实验拟合曲线与模型训练输出曲线重合性较好,表现出模型较强的预测能力,再通过训练误差曲线可以看出,当载荷较小时,训练输出的法向接触刚度较实际值误差较大,这是因为开始训练的样本参数较少,不能准确输出法向接触刚度,但随着法向载荷的增大,误差急剧减小,预测能力显著提高,训练误差 0.003%,并且误差趋于恒定,误差方差值为6.43e－10,表现出误差波动较小。

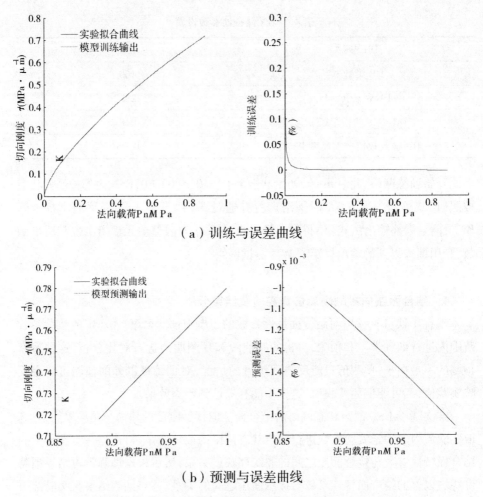

图2 实验1条件下法向接触刚度的训练与预测曲线

图2(b)是在图2(a)训练的基础上预测接触载荷在 0.9－1MPa 范围内的法向接触刚度值,由预测曲线和实验曲线对比可以看出,两曲线重合度较好,没有出

现震荡现象,从预测误差曲线进一步可以看出,预测误差虽然随着法向载荷的增大而增大,但是预测精度较高,最大误差为 -0.002%,预测误差方差为 2.34e-11,进一步说明采用 ESN 回声网络算法能有效预测非线性特性问题。

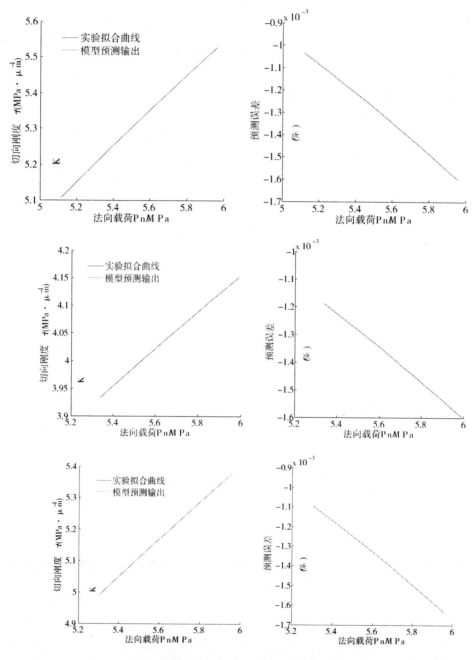

图3 实验2、3、4条件下预测与实验曲线

限于篇幅,图3分别给出了在表3中2、3、4组合条件下的法向接触刚度的预测与实验对比曲线和预测误差曲线,由各组曲线可以看出,预测曲线和实验曲线一致性较好,预测精度较高,预测误差较小,但是都表现出随着法向接触载荷的增大,预测误差也随着增大,因此,可以得出如果载荷过大,预测值会偏离实验值。从另一方面讲,当结合面接触载荷过大时,材料已经严重发生塑性变形,此时机械构件已经失效,无论是理论计算还是实验研究失去其意义。所以,由建模和计算结果表明,该算法能有效预测结合面法向接触刚度。

4.2 预测能力评价

根据表3几种不同组合条件下组合,表5列出了基于ESN回声网络算法结果的最大预测误差和预测方差值,从表中可以看出,在特定的预测域内,预测的相对误差基本保持在0.0016%以内,预测能力基本保持稳定,再由预测方差结果表明,预测曲线和实验曲线一致性较好,预测曲线较实验曲线的波动性很小。

表5 结合面法向接触刚度的预测误差和方差

序号	配对副材料	最大误差(%)	预测方差
1	铸铁 – 贴塑	– 0.0016	2.34e – 11
2	45钢 – 铸铁	– 0.0016	6.58e – 12
3	铸铁 – 铸铁	– 0.0016	1.37e – 11
4	铸铁 – 铸铁	– 0.0017	7.47e – 12

4.3 不同算法预测能力对比分析

为了进一步评价ESN回声网络算法预测结合面法向接触刚度的预测能力,采用本文表3中的第4种条件,隐含层为8层,迭代次数为1000次,对本文提出的结合面法向接触刚度的ESN算法与BP神经网络、RBP神经网络、MPSO – BP网络算法进行对比,最大误差和预测方差的值如表6所列,由表6中的对比数据可以看出,传统BP神经网络的预测能力最低,ESN回声网络算法的预测能力最高,这与文献[15 – 18]中的其它领域预测性能对比结果相一致。

表6 结合面法向接触刚度的预测性能对比

序号	算法类别	最大误差(%)	预测方差
1	BP 神经网络	9.78	1.85e−2
2	RBP 神经网络	−7.21	8.53e−2
3	MPSO−BP 网络	5.01	3.37e−3
4	ESN	−0.0017	7.47e−12

5 结合面法向接触刚度在机床中的应用

机床中大量存在大量的结合面,而零件与结合面既互相依附,又互相影响,机床结构的结合面接触特性对机床整机的分析有着较大的影响,主要的表现形式是机床工作时的变形[19]。本文以某机床企业生产的 CK61125 的床身导轨副为研究对象,基于吉村允孝法利用本文预测的结合面法向特性参数,对其变形进行有限元的计算和实验采集。实验导轨副结合面材料为 HT200,外载荷为 5000N,材料泊松比为 0.26,实验与计算结果如表 7 所列。

表7 导轨副接触变形计算和实验对比

采样点	理论变性(μm)	实测变性(μm)	相对误差(%)
1	−11.3	−10.4	−8.0
2	−8.3	−7.6	−8.4
3	3.8	4.1	−7.9
4	8.5	8.1	4.7

由表 6 中可以看出,采样点的理论计算和实验结果基本接近,相对误差值在 10% 之内,误差较大主要原因是用理论计算的方法不能准确模拟实际结构的边界条件,用有限元方法进行建模时进行了等效方法处理,造成了一定的计算误差,同时,实测数据也是由实验测得,从其实验本身而言存在一定的误差。结果表明,该方法对工程应用具有一定的理论指导意义。

6 结论

本文提出了基于 ESN 回声状态网络算法对结合面法向接触刚度进行建模和预测,构建了训练学习域值与预测域值相分离的理念,通过 4 种不同组合条件下的结合面接触刚度的参数设置、误差分析和方差值分析,结果表明,该算法对本文算例的最大误差小于0.0016%,具有较好的预测能力。同时在一定组合条件下,对其它算法预测性能进行对比,也获得更高精度的辨识能力,并将其应用到机床导轨副的接触变形解析中。因此,本文提供的方法是一种较实用的非线性系统辨识方法,进一步在结合面动态接触特性中加以应用,值得在工程领域推广应用。

参考文献

[1] KONOWALSKI K. Experimental research and modeling of normal contact stiffness and contact damping of machined joint surfaces[J]. Advances in Manufacturing Science and Technology,2009,33(3):53-68.

[2] 杨红平,傅卫平,王雯,等.基于分形几何与接触理论的结合面法向接触刚度计算模型[J].机械工程学报,2013,49(1):102-107.
YANG H P, FU W P, WANG W, et al. Calculation model of the normal contact stiffness of joints based on the fractal geometry and contact theory[J]. Journal of Mechanical Engineering, 2013,49(1):102-107.

[3] JENG L L,JEN F L. A modified fractal microcontact model developed for asperity heights withvariable morphology parameters[J]. Wear,2010,(268):133-14.

[4] GREENWOOD J A,WILLIAMSON J B P. Contact of Nominally Flat Surfaces[J]. Proc. Roy. Soc. ,London A295,1966:300-319.

[5] MAJUMDAR A, BHUSHAN B. Fractal model of elastic-plastic contact between rough surfaces[J]. ASME Journal of Tribology,1991,113:1-11.

[6] 尤晋闽,陈天宁.基于分形接触理论的结合面法向接触参数预估[J].西安交通大学学报,2011,45(9):1275-1280.
YOU J M, CHEN T N. Estimation for normal parameters of joint surfaces based on fractal theory[J]. Journal of Shanghai Jiaotong University. 2011,45(9):1275-1280. PH

[7] JIANG SH Y, ZHENG Y J, ZHU H. A contact stiffness model of machined plane joint based on fractal theory[J]. Journal of Tribology, ASME,2010,132:011401-7.

[8] SONG W P,LI L Q,ETSION I,et al. Yield inception of a soft coating on a flat substrate indented by a rigid sphere[J]. Surface & Coatings Technology 2014,240:444-449.

[9] FU W P, HUANG Y M, ZHANG X L. Experimental Investigation of Dynamic Normal Characteristics of Machined Joint Surfaces[J]. Journal of Vibration and Acoustics,2000,122(4):393-398.

[10] PITENIS A A, DOWSON D, SAWYER W G. Leonardo da Vinci's Friction experiments: an old story acknowledged and repeated[J]. Tribol. Lett. 2014,56:509-515.

[11] SONBATY I A, KHASHABA U A, SELMY A I, et al. Prediction of surface roughness profiles for milled surfaces using an artificial neural network and fractal geometry approach[J]. Journal

of Materials Processing Technology,2008,200:271-278.

[12] 温淑花,张学良,倪润堂.机械结合面切向接触阻尼的神经网络结构化建模[J].农业机械学报,2002,33(1):87-89.

WEN SH H,ZHANG X L,NI R T. Modeling of tangential contact damping in machine joints using neural network[J]. Transactions of the Chinese Society for Agricultural Machinery,2002,33 (1):87-89.

[13] 杨红平,傅卫平,王雯,等.结合面法向和切向接触刚度的 MPSO-BP 神经网络算法的建模.仪器仪表学报,2012,33(8):1856-1861.

YANG H P,FU W P,WANG W,et al. Machine joints normal and tangential contact stiffness modeling based on MPSO-BP neural network algorithm[J]. Chinese Journal of Scientific Instrument,2012,33(8):1856-1861.

[14] 朱坚民,张统超,李孝茹,等.基于改进自适应遗传算法的固定结合面动态特性参数优化识别[J].中国机械工程,2014,25(3):357-365.

ZHU J M,ZHANG T CH,LI X R,et al. Optimiztion identification for dynamic chracteristics parameters of fixed joints based on improved adaptive genetic algorithm[J]. China Mechanical Engineering,2014,25(3):357-365.

[15] 李晓华,李军.基于 ESN 网络的连续搅拌反应釜(CSTR)辨识[J].信息与控制,2014,43(2):223-228.

LI X H,LI J. Identification of continuous stirred tank reactor based on echo state network[J]. Information and Control,2014,43(2):223-228.

[16] 彭宇,王建民,彭喜元.基于回声状态网络的时间序列预测方法研究[J].电子学报,2010,38(2A):148-154.

PENG Y,WANG J M,et al. Researches on time series prediction with echo state networks[J]. Journal of Electronics,2010,8(2A):148-154.

[17] JAEGER H,HASS H. Harnessing nonlinearity:predicting chaotic systems and saving energy in wireless communication[J]. Science,2004,304(5667):78-80.

[18] 雷苗,彭宇,郭嘉,等.基于先验簇复杂回声状态网络的话务量预测[J].仪器仪表学报,2011,32(10):2190-2196.

LEI M,PENG Y,GUO,J,et al. Traffic forecasting for prior knowledge based clustered complex echo state networks[J]. Chinese Journal of Scientific Instrument,2011,32(10):2190-2196.

[19] HE T,REN N,ZHU D,et al. Plasto-elastohydrodynamic lubrication in point contacts for surfaces with three dimensional sinusoidal waviness and real machined roughness[J]. Journal of Tribology,ASME,2014,136/031504-1.

注:本文曾发表在 2016 年《仪器仪表学报》第 4 期上。

Investigation of Tool Wear and Surface Roughness when Turning Titanium Alloy (Ti6Al4V) under Different Cooling and Lubrication Conditions

Limin Shi *

摘 要:钛合金(Ti6Al4V)切削时效率低、加工表面质量差,因此,用CNMG120408 刀具在三种不同的冷却和润滑条件(干式切削、润滑切削和MQL)下对钛合金(Ti6Al4V)进行了切削实验,通过实验,对在三种不同的冷却和切削条件下刀具磨损和表面粗糙度的影响进行了分析,干式切削中,前刀面磨损变化不大,而润滑切削和MQL对刀面磨损有明显影响。切削液使切屑的锯齿化程度增加,影响切削力,从而影响了表面质量。

Because of the low efficiency and low machined surface integrity during cutting Ti6Al4V, the turning experiments of Ti6Al4V alloy were carried out with ultra-fine grain coated carbide tools, CNMG120408 under three different cooling and lubrication conditions, namely dry cutting, wet cutting and Minimum Quantity Lubrication(MQL). The influence of cooling and lubrication strategies on tool wear and surface roughness were analyzed. Compared with dry cutting, the rake face wear had little change, while significant difference of the flank face wear had been observed during wet cutting and MQL. As a result of the increase of the degree of chip serration caused by cutting fluid, the fluctuation of cutting force occurred, and then the surface roughness had been influenced.

---

* 作者简介:时立民(1978—),男,甘肃清水人,天水师范学院副教授、硕士,主要从事金属切削技术和现代制造技术领域的研究。

Introduction

Ti6Al4V are used in a wide range of applications in the aerospace, automotive, chemical and medical industries due to their low thermal conductivity coefficient, small elastic modulus and high chemical activity. However, the high cutting force, excessive tool wear and low surface quality, resulting in lower production efficiency. Therefore, reducing tool wear and improving surface quality are the necessary considerations for improving tool life and productivity.

To ensure the high productivity and good surface quality while reducing the costs, the different cooling lubrication strategies is an obvious progress. MQL is undoubtedly the most economic one. In the device of MQL, the coolant, generally vegetable oil is mixed with the compressed air flow and then is jetted into the cutting area[1]. Feng Jiang [2] performed cutting tests under conditions of dry cutting, wet cutting and MQL. Zhou [3] applied this technology into the machining operation of titanium alloy. The related researches on the optimization of cutting parameters with MQL were carried out by Zhi Qiang Liu during titanium alloy process [4]. In addition, Klocke [5] evaluated tool wear, cutting forces, chip morphology and surface roughness during MQL. However, most of the researches mentioned above are divided, only concerned on one or a few phenomena occurred in machining process. This paper will analyze the effects of MQL on titanium alloy machining systematically including cutting force, breakage and wear of cutting tool on either rake or flank surface, machined surface roughness and compare the effect of different cooling and lubrication strategies.

This paper is structured as follows: In the first section, the tool CNMG120408 is selected to use in the cutting experiments with different conditions. In the second and third section, the tool wear on either rake and flank face is contrasted and analyzed each other, while in the fourth sections for this study, the machined surface roughness are researched, respectively. Finally, there is a conclusion section.

1. Experimental scheme and conditions

Ti6Al4V bar was chosen as the workpiece material, whose chemical composition is as follows, 0.1% of C, 0.3% of Fe, 0.05% of N, 0.02% of O, 5.5 ~ 6.8% of Al, 3.5 ~ 4.5% of V, 0.015% of H, and Ti of the rest. The yield strength and tensile strength are about 825MPa and 910MPa, respectively. Its elastic modulus is approximately 118GPa, and elongation rate is 10%. CA6150 CNC lathe was applied to these cylindrical turning experiments, where Mitsubishi (Japan) ultrafine grain carbide coated tools were adopted with the code CNMG120408 – MJ.

The turning experiments of Ti6Al4V alloy were carried out by CNMG120408 with differentcooling and lubricating strategies where the cooling conditions are air cooling (dry), conventional flood cooling (wet), and MQL. In MQL, the LUBROIL lubricating oil, a kind of fatty acid ester, is used as the lubricant and its flow rate is set to 150mL/h. And the contrast cutting test each other was done in order to the research the influence of cooling conditions on cutting process. After turning tests, the wear morphologies of the tools used were observed by optical microscope, the machined surface roughness was measured. The test setup and parameters were given in table 1.

Table 1 Test setup and parameters

Condition / Cutting tool	CNMG120408 – MJ		
Cooling condition	Dry	Wet	MQL
Running time	15min		
Cutting speed: V_c	40m/min, 80m/min, 120m/min		
Feed value: f	0.1mm/r, 0.15 mm/r, 0.2 mm/r		
Depth of cut: a_p	0.5mm		

2. Tool Wear on rake face under different cooling and lubrication conditions

Rake face wear state under three kinds of cutting conditions was shown in Fig. 1, the degree of wear which mainly reflects in the wear depth is positive correlated with cutting speed at arbitrarily defined cooling conditions, this relates to higher cutting temperature during larger cutting speed. However, from the view of the area of wear, there is no obvious change in rake face wear under different cooling and lubrication conditions. This is because that cutting fluid cannot enter the tool – chip contact area, as a result that it cannot play a role of cooling and lubrication. So the wear state on rake face has not been well improved.

In addition, it can also be found that each of the rake face wears with different cutting parameters and cooling strategies is mainly manifested as crater wear. The chief rea-

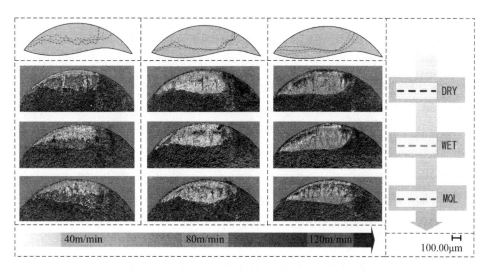

Fig. 1 Rake face wear state under three kinds of cutting conditions ($f = 0.15$mm/r, $a_p = 0.5$mm)

son is that the high-temperature chemical activity and chemical affinity of titanium alloy is high [6], leading to adhesion easily with the tool in cutting process. Along with machining, the frictional interaction between tool and chip will make the adhesive material falling off gradually, which may take part of the tool surface material away. As a result, the tool-chip interface maintains fresh at all times, and then the oxidation and diffusion are sharpened. This may reduce the anti-shear ability and external hardness of the tool material, and finally causing the crater wear phenomenon described above (see Fig. 2). So the cutting performance of the tool is weakened, and the failure is induced later.

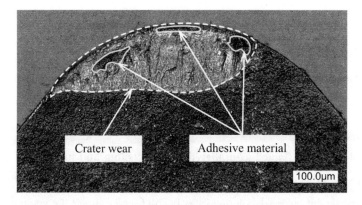

Fig. 2 Main wear form of rake face (dry cutting, $v_c = 120$m/min, $f = 0.15$mm/r, $a_p = 0.5$mm)

To obtain the formation mechanism of crater wear in the cutting process of titanium alloy and to verify the existence of diffusion effect between the tool and workpiece on rake face, EDS analysis of the tool body and the adhesive material on rake face were carried out. The element distribution diagram of rake face was got as shown in Fig. 3 (point analysis). It can be seen from Fig. 3a that, C and W element contained in tool body are found in the adhesive material which formed by workpiece material. This illustrates that the elements in the cutting tool diffuse outward in the machining process. In addition, higher content of Ti exists in the tool body, as shown in Fig. 3b, which is because of Ti contained in the coating material. So it cannot be determined whether the Ti is from the workpiece or not. But the V only contained in the workpiece is also found in the tool body, which just illustrates that the elements in the workpiece diffuse into the tool body. With respect to a and b in Fig 3, it can be concluded that there is element diffusion between tool and workpiece, and no good inhibition effect produced by cutting fluid on oxidation and diffusion is observed through the energy spectrum analysis.

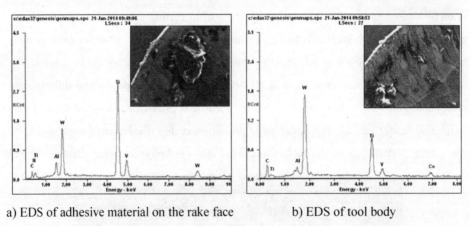

a) EDS of adhesive material on the rake face   b) EDS of tool body

Fig. 3 Element analysis on rake face (dry cutting, $v_c = 120\text{m/min}$, $f = 0.15\text{mm/r}$, $a_p = 0.5\text{mm}$)

3. Tool Wear on flank face under different cooling and lubrication conditions

Similar to the rake face wear, the flank wear also becomes serious with the increase of cutting speed. But obviously suppress has been observed under the cooling and lubrication conditions, as shown in Fig. 4. Compared the three kinds of cooling and lubricating condition, it can be seen that the degree of flank wear has been obviously controlled under the cutting fluid condition when the cutting speed is 80m/min and 120m/min. The reason is that the effect of cooling and lubrication reduces the cutting temperature and then leads to the decrease of adhesive material and finally inhibits the adhesive wear effectively. The cutting edges obviously deform during dry cutting when the cutting speed is 120m/min, while the deformation of the cutting edge becomes small with pouring cutting fluid, and even there is almost no deformation on the cutting edge under the MQL condition. Based on the analysis above, it is known that the cooling and lubrication effect of the cutting fluid effectively reduces the degree of tool wear and oxidation, and MQL is a promising cooling method.

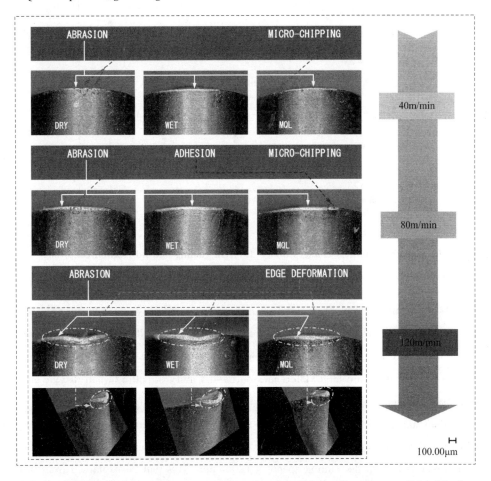

Fig. 4　Flank and cutting edge wear state under three kinds of cutting condition($f = 0.15$mm/r, $a_p = 0.5$mm)

4. Influence of different cooling and lubrication conditions on machined surface roughness

Machined surface roughness is seen as an important index of machining performance, which is comprehensively reflected by the influence of cutting parameters and variables on cutting process in machining system. Fig. 5 shows the change trend of surface roughness value under different cooling and lubrication conditions.

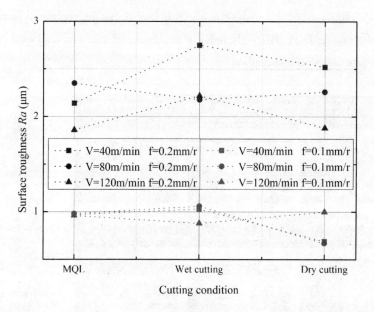

Fig. 5 Surface roughness changes under different cooling and lubrication conditions ($a_p = 0.5$mm)

As shown in Fig. 5, the machined surface roughness value becomes higher obviously with the increase of feed, which illustrates that feed has a significant effect on the surface roughness. Cutting speed has little influence on the surface roughness under MQL condition, while there are large differences during dry and wet cutting when varying cutting speed, especially in dry cutting. The surface roughness obtained from dry cutting is lower than that in wet cutting and under MQL conditions when the feed value is 0.1mm/r and the cutting speed is 40m/min or 80m/min. However, the opposite conclusion is received when the cutting speed is 120m/min. This can be explained from two aspects. on the one hand, the serrated chip will be produced in the cutting process of titanium alloy[7], and the cutting fluid makes the degree of chip serration increasing,

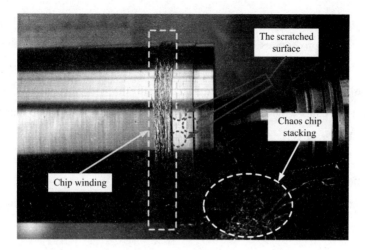

Fig. 6 The chip winding phenomenon on workpiece
($v_c = 80\text{m/min}, f = 0.15\text{mm/r}, a_p = 0.5\text{mm}$)

which will cause the cutting force fluctuating more seriously [8], and then result that the surface roughness values under MQL and wet cutting conditions are higher than that during dry cutting. On the other hand, the tool wear is not obvious at the speed of 40m/min and 80m/min. But when the cutting speed increases to 120 m/min, not only the cutting temperature turns high and the tool wear becomes serious, but also the phenomenon that chips discharged hardly stack on the tool tip and wind with the workpiece will occur. The behavior of continuously scratching workpiece by chips causes the surface quality to decline, as shown in Fig. 6. In addition, some literatures show that tool wear plays an important role on changing the surface roughness when the cutting parameters and cooling conditions are fixed.

5. Conclusion

1) Compared with dry cutting, the rake face wear of CNMG120408 is not effectively improved during wet cutting and MQL. The tool wear on rake face is crater wear during turning Ti6Al4V in three different cooling and lubrication conditions and its main formation mechanism are adhesion and diffusion.

2) The flank face wear is evidently restrained under the cooling and lubrication conditions. Especially in cutting speed of 120m/min, significant deformation has been produced on the cutting edge of CNMG120408 during dry cutting, but the deformation is small under conventional cooling, even the deformation almost disappears with MQL.

3) In the cutting speed of 40 m/min and 80 m/min, the degree of chip serration turns more and more serious due to cutting fluid, which induces the large fluctuation of cutting force, thus the surface roughness is made to be larger than that during dry cutting. But when cutting with a speed of 120 m/min, the tool wear of CNMG120408 is drastic and the chips wind around the workpiece, meanwhile, the surface roughness during dry cutting is larger.

Acknowledgements

This research is supported by the National Natural Science Foundation of China (51665053).

References

[1] Yan Lutao, Yuan Songmei, Liu Qiang. Tool Wear and Chip Formation in Green Machining of High Strength Steel. Journal of Mechanical Engineering 2010,46(9):187 - 192

[2] Feng Jiang, Jianfeng Li, Lan Yan, Jie Sun, etc. Optimizing end - milling parameters for surface roughness under different cooling/lubrication conditions. Int J Adv Manuf Technol 2010,51(9 - 12):841 - 851

[3] Zhou Zhi - min, Zhang Yuan - liang, Li Xiao - yan, Zhou Hui - yuan. Research on application of gas - liquid atomization technology to cutting titanium alloy workpiece. Journal of Dalian University of Technology 2011,51(1):41 - 45

[4] ZhiQiang Liu, Ming Chen, QingLong. AnInvestigation of friction in end - milling of Ti - 6Al - 4V under different green cutting conditions. Int J Adv Manuf Technol 2015,78(5 - 8):1181 - 1192

[5] Fritz Klocke, Luca Settineri, Dieter Lung, Paolo Claudio Priarone, et al. High performance cutting of gamma titanium aluminides: Influence of lubricoolant strategy on tool wear and surface integrity. Wear 2013,302(1 - 2):1136 - 1144

[6] Kyung - Hee Park, Gi - Dong Yang, M. A. Suhaimi, Dong Yoon Lee, etc. The effect of cryogenic cooling and minimum quantity lubrication on end milling of titanium alloy Ti - 6Al - 4V. Journal of Mechanical Science and Technology,2015,29(12):5121 - 5126

[7] G. Sutter, G. List. Very high speed cutting of Ti - 6Al - 4V titanium alloy - change in morphology and mechanism of chip formation. International Journal of Machine Tools & Manufacture 2013,66:37 - 43

[8] S. Sun, M. Brandt, M. S. Dargusch. Characteristics of cutting forces and chip formation in machining of titanium alloys. International Journal of Machine Tools & Manufacture 2009,49(7 - 8):561 - 568

注:本文曾发表在2018年《FERROELECTRICS》第1期上

基于小波包能量曲率差的古木结构损伤识别

王 鑫　胡卫兵　孟昭博*

以西安钟楼为工程依托,对随机激励作用下古木结构的损伤进行有限元模拟,把古木结构梁上各节点的加速度响应信号进行小波包分解,通过小波包能量曲率差对古木结构进行损伤定位。在无噪声干扰时,该指标对于古木结构的损伤定位比较敏感,可准确判定古木结构的损伤位置,该指标随损伤程度的加大而增大。该指标在高斯白噪声干扰下,当信噪比 SNR 大于或等于40db 时,能对古木结构的损伤进行准确定位,该指标具有一定的抗噪声干扰能力。随后得出了损伤指标和损伤程度之间的函数关系式,用其进行损伤程度的判断并验算其适用性,为研究环境激励下西安钟楼的损伤预警提供了理论依据。

我国是文明古国,古建筑是重要的历史文化遗产组成部分,是中华民族的瑰宝。西安钟楼位于东、西、南、北四条大街交汇处,是西安的标志性建筑,是国家重点保护文物。上世纪80年代地面交通引起的微幅振动使钟楼二层地板上的展品发生移位,高台基的墙体和木构架的横梁裂缝不断扩展。虽然钟楼环道内侧至高台基边缘的距离扩大到26.7m,车辆引起的振动对钟楼的影响有增无减,钟楼二层地板有明显的振感。西安是西部发展重心,现代工业迅速发展,人口不断增加,地面交通流量不断增大,为从根本上缓解地面交通压力,西安市规划建设6条地铁线路,总长达251.8km,钟楼四侧有四条隧道穿过,距离钟楼仅15m。对古木结构来说,交通环境振动是历经数百年古建木结构寿命新威胁,也是正常情况下经受的最严重的威胁,长期的交通环境微振动能使木结构疲劳、榫卯松动、结构变形、寿命缩短。钟楼一旦遭到破坏,不可再生,造成历史文化遗产不可弥补的损失,这就要求结构健康监测与安全评估系统及时发现损伤并在第一时间预警。因

---

* 作者简介:王鑫(1971—),女,陕西西安人,天水师范学院副教授,博士,主要从事结构健康监测研究。

此利用环境激励下的结构振动响应对结构进行健康监测,已是结构健康监测中损伤检测的新研究方向,同时对保护古建筑有着积极意义。

通常的损伤识别系统分为四个阶段:1)损伤是否存在;2)损伤位置判定;3)损伤的严重程度;4)结构的实用性预测。损伤识别方法分为局部和整体两种方法。目前有直观判定、超声波、电磁场及电涡流等检测方法,需预先知道损伤的大体位置,对损伤部位进行仪器操作,整体损伤识别通过结构振动特性的变化评价结构的健康情况,只有将两者结合起来,才能准确评价复杂结构的健康状况[1-2]。传统的模态参数频率和振型不能准确识别损伤,大多数损伤评估方法通过对瞬态信号的傅立叶变换得到模态参数,而傅立叶变换的最大缺陷是对高频模态分析不足。小波分析被誉为数学显微镜,本身具有放大、缩小和平移等功能,可通过检查不同放大倍数下的变化来研究信号的特征,具有优良的时频局部化特性,但有高频段分辨率差的缺点。小波包分析是小波变换的扩充,能为信号提供一种更加精细的分析方法,将小波分析没有细分的高频部分进一步分解,具有任意的时-频分辨率,根据分析信号的特征选择相应的频带,使之与信号频谱匹配。因此小波包分析在土木工程结构的健康监测与损伤诊断中具有非常广阔的应用前景。

20世纪80年代开始,国内外学者开展了小波包分析在结构损伤诊断领域中的应用研究,取得了一定的研究成果。丁幼亮等[3]对Benchmark钢框架结构试验数据和润扬大桥悬索桥监测数据进行小波包能量谱损伤预警分析,在此基础上详细考察了不同小波函数和小波包分解层次的损伤预警效果。刘涛等[4]做了基于小波包能量谱的结构损伤预警方法试验研究。邓扬等[5]采用小波包分析进行了拉索损伤声发射信号特征提取。范颖芳等[6]利用小波包对不同损伤情况下拱桥结构的观测信号分析,确定了结构异常状态的敏感特征。韩建刚等[7]提出了小波包变换的能量变化率指标对梁体进行了损伤识别的定位研究,并进行试验验证。余竹等[8]利用沧州子牙河新桥替换下的梁体进行两种工况损伤模拟,用小波包能量曲率差法识别损伤,考察小波函数和分解层数对识别效果的影响。

以西安钟楼为工程依托,提出小波包能量曲率差对古木结构进行损伤定位识别,把有限元分析得到梁上各节点的加速度响应信号进行小波包分解,计算该指标进行损伤定位,该指标对于损伤识别比较敏感,能准确判定古木结构损伤的具体位置。

1 小波包分析

小波包由一系列线性组合小波函数组成:

$$\psi_{j,k}^{i}(t) = 2^{\frac{j}{2}}\psi^{i}(2^{j}t - k) \quad i = 1, 2, \cdots \tag{1}$$

式中，i、j、k 分别表示频率因子、尺度因子和平移因子。

小波函数 ψ^i 递推关系式为：

$$\psi^{2i}(t) = \sqrt{2}\sum_{k=-\infty}^{\infty}h(k)\psi^{i}(2t - k) \tag{2}$$

$$\psi^{2i+1}(t) = \sqrt{2}\sum_{k=-\infty}^{\infty}g(k)\psi^{i}(2t - k) \tag{3}$$

式中：ψ 表示小波母函数，h(k)和g(k)为与尺度函数及小波母函数相关的积分镜像滤波器系数。

对于任意信号的第 j 阶和第 j+1 阶水平小波包分解递推关系为：

$$f_j^i(t) = f_{j+1}^{2i-1}(t) + f_{j+1}^{2i}(t) \tag{4}$$

$$f_{j+1}^{2i-1}(t) = Hf_j^i(t) \tag{5}$$

$$f_{j+1}^{2i}(t) = Gf_j^i(t) \tag{6}$$

其中，H 和 G 分别为 h(k)和 g(k)构成的滤波算子，

$$H\{\bullet\} = \sum_{k=-\infty}^{\infty}h(k-2t) \tag{7}$$

$$G\{\bullet\} = \sum_{k=-\infty}^{\infty}g(k-2t) \tag{8}$$

经过 j 水平的小波包分解后，初始信号 f(t)为：

$$f(t) = \sum_{i=1}^{2^j}f_j^i(t) \tag{9}$$

小波包组分信号 $f_j^i(t)$ 表示为小波包函数的线性组合：

$$f_j^i(t) = \sum_{-\infty}^{\infty}c_{j,k}^i(t)\psi_{j,k}^i(t) \tag{10}$$

小波包系数为：

$$c_{j,k}^i = \int_{-\infty}^{\infty}f(t)\psi_{j,k}^i(t)dt \tag{11}$$

小波包系数满足正交条件：

$$\psi_{j,k}^m(t)\psi_{j,k}^n(t) = 0 \quad (m \neq n) \tag{12}$$

小波包信号能量为：

$$E_f = \int_{-\infty}^{\infty}f^2(t)dt = \sum_{m=1}^{2^i}\sum_{n=1}^{2^i}\int_{-\infty}^{\infty}f_j^m(t)f_j^n(t)dt \tag{13}$$

将式(10)代入式(13)，并利用式(12)得到：

$$E_f = \sum_{i=1}^{2^j}E_{f_j} \tag{14}$$

小波包组分能量 E_{f_j} 视为存储于组分信号 $f_j(t)$ 的能量：

$$E_{f_j} = \int_{-\infty}^{\infty} f_j(t)^2 dt \tag{15}$$

小波包组分能量对信号的变化十分敏感，可用于结构的损伤识别。

2 古木结构的损伤识别

把古木结构的损伤识别分为两部分：(1)对古木结构梁上各节点的加速度响应信号进行小波包分解；(2)计算小波包能量曲率差进行结构的损伤定位识别，其中包括小波函数和小波包分解层次选择。

2.1 合理的选择小波函数

从消失矩和支撑长度考虑，选用 Daubechies 为小波函数，简记为 dbN(N 为阶次)。N 越大，Daubechies 小波的消失矩越高，时域的分辨率越好；但同时 Daubechies 小波支撑长度越宽，小波的时域局域性越差。因此应合理确定 Daubechies 小波阶次 N。

对结构动力响应 $f(N,k)$ 进行第 i 层小波包分解，f_{ij} 表示第 i 层分解节点 (i,j) 的结构响应，每个频带内结构响应 f_{ij} 能量[9]：

$$E_{ij} = \sum |f_{ij}|^2 \quad (j=0,1,2,\cdots,2^i-1) \tag{16}$$

则结构动力响应 $f(N,k)$ 第 i 分解层的小波包能量谱向量 E_i：

$$E_i = \{E_{i,j}\} = [E_{i0}\ E_{i1}\cdots\ E_{ij}\cdots\ E_{i2^i-1}]^T \tag{17}$$

为衡量小波函数好坏，定义 i 分解层各频带能量系数系列 $\{E_{ij}\}$ 的代价函数 $M\{E_{ij}\}$。小波包能量谱中各频带能量系数 E_{ij} 的时频集中程度由代价函数 $M\{E_{ij}\}$ 反映。采用 l^p 范数熵为代价函数，在同一小波包分解层上，计算不同小波函数的代价函数值并比较，确定较适合的 Daubechies 小波阶次 N。通常不同的阶次计算小波函数的代价函数值越小越好，l^p 范数熵 $(1 \leqslant p \leqslant 2)$ 定义为[9]：

$$S_L(E_i) = \sum_j |E_{i,j}|^p \tag{18}$$

2.2 合理的选择小波包分解层数

在工程应用中，对结构动力响应进行小波包分解，计算每一分解层次上的小波包能量谱的代价函数，从代价函数和计算时间考虑确定适当的小波包分解层次，通常小波包能量谱代价函数值越小，计算机计算过程耗时越少，小波包分解层次越好。类似小波函数阶次的选择方法，采用 l^p 范数熵为代价函数，确定合适的小波包分解层数，l^p 范数熵 $(1 \leqslant p \leqslant 2)$ 定义[9]：

$$S_L(E_i) = \sum_j |E_{i,j}|^p \tag{19}$$

2.3 不等间距曲率求解方法[8]

在实际工程中,曲率一般由变量的二阶差分(斜率的变化率)得到。

不等间距情况下曲率求解式为:

$$K_i = y_i^* = \frac{\dfrac{y_{i+1} - y_i}{h_{i+1}} - \dfrac{y_i - y_{i-1}}{h_{i-1}}}{\dfrac{h_{i+1} + h_{i-1}}{2}} \tag{20}$$

式中:分子为节点左右两段曲线斜率差,分母为节点左右两端斜率差间距。若节点等间距,$h_{i-1} = h_{i+1}$,则

$$K_i = y_i^* = \frac{\dfrac{y_{i+1} - y_i}{h} - \dfrac{y_i - y_{i-1}}{h}}{h} = \frac{y_{i+1} - 2y_i + y_{i-1}}{h^2} \tag{21}$$

式(21)为二阶差分法求解等间距曲率公式。将式(21)中 y 换成小波包能量谱,则为小波包能量曲率。将完好状态与损伤状态各节点的小波包能量曲率进行插值,得到损伤状态的小波包能量曲率差为:

$$\Delta K_i = K_i^u - K_i^d \tag{22}$$

式中:k_i^u,k_i^d 分别为完好状态与损伤状态的小波包能量曲率。

3 算例

3.1 古木结构的有限元模拟

本文以西安钟楼为工程依托,选取其中一榀框架进行分析。通过 Ansys 有限元软件对环境激励下古木结构进行损伤模拟分析,以钟楼为参考,选取木框架计算参数,选取木梁长4m,木柱高6m,梁截面尺寸为 $300 \times 700 \text{ mm}^2$,柱截面直径500 mm,用 beam188 梁单元模拟木柱、木梁,用 combin14 单元模拟榫卯节点,榫卯连接的弯曲刚度为[10]:$1 \times 10^{10} \text{kN} \cdot \text{m/rad}$,木材的弹性模量取 $1 \times 10^{10} \text{N/m}^2$,泊松比为0.25,密度为410kg/m³,采用 Rayleigh 定义的粘性比例阻尼。柱子搁置在有凹槽的柱础上,不能完全限制柱子转动,在荷载引起的微幅振动下不计其线位移,柱与基础的连接简化成固定铰支座的力学模型符合实际情况[11-12],建立古木结构的有限元模型如图1所示。

西安钟楼处于地面交通和地铁运行的复杂交通环境下,地面交通振动通过高台基传播引起钟楼振动,运行的地铁对轨道产生的冲击作用产生振动,通过隧道结构传到周围地层,并经过地层向周围传播激励钟楼产生振动,因而在该木框架的柱底节点 1 处沿 x 轴正方向施加随机激励荷载来模拟环境激励对钟楼的影

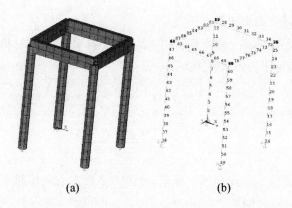

图1 古木结构的有限元模型(a)和节点详图(b)

响[13],随机激励荷载的时程曲线及频谱如图2所示,获得结构的加速度荷载时程,在此基础上运用Matlab程序计算了小波包能量谱。

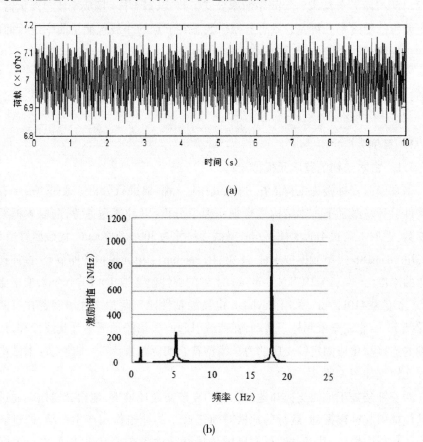

图2 激励荷载的时程曲线(a)和频谱曲线(b)

古木结构的损伤程度通过折减损伤单元的弹性模量来实现,其中10%、18%、20%分别指损伤单元的弹性模量减少10%、18%、20%,如表1所示[9]。

对损伤工况1、2进行分析,得出完好结构和损伤工况1、2梁跨中第31节点的竖向加速度时程曲线如图3所示。从图3看出各损伤工况的信号有细微差别,但很难判断古木结构的损伤情况。因此下面采用小波包能量曲率差对古木结构进行损伤识别。

表1 古木结构损伤工况

损伤工况	损伤位置	损伤 程度/%	说明
1	第52、第53单元 (对应节点30、31、32)	10	梁跨中附近
2	第52、第53单元 (对应节点30、31、32)	20	梁跨中附近
3	第52、第53单元 (对应节点30、31、32)	18	梁跨中附近

3.2 选择计算参数

3.2.1 小波函数的选择

梁跨中挠度较大,是易出现损伤部位,因而对完好结构梁跨中第31节点的竖向加速度响应,选择不同阶次的Daubechies进行小波包分解,分解层次取4,计算l^p范数熵代价函数值如表2所示。从表2看出,当小波阶次为20时,l^p范数熵为5592.46,其值相对其他小波阶次最小,因此损伤识别小波函数选择Daubechies20。

表2 分解层次为4时不同db小波的代价函数值

小波阶数 N	5	6	7	8
$S_1(E_j)(p=1.5)$	6885.73	6833.38	8863.13	6057.81
小波阶数 N	9	10	11	12
$S_1(E_j)(p=1.5)$	7984.79	6672.64	10700.35	8525.24
小波阶数 N	13	14	15	16
$S_1(E_j)(p=1.5)$	6991.40	6060.01	7343.79	8976.83
小波阶数 N	17	18	19	20
$S_1(E_j)(p=1.5)$	7856.51	8733.06	6595.94	5592.46

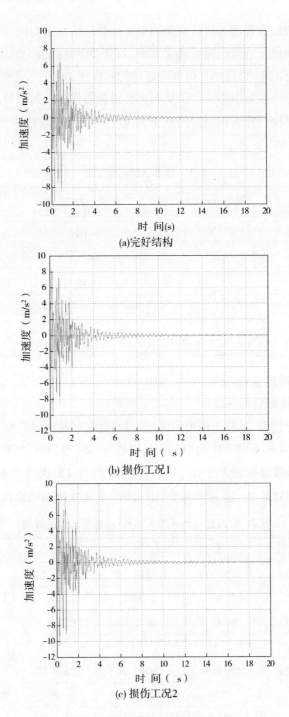

图3 第31节点加速度响应

3.2.2 小波包分解层数的选择

采用 Daubechies20 对梁跨中第 31 节点完好状态下的竖向加速度响应进行小波包分解,分解层次取 1~8,计算 l^p 范数熵的代价函数值,并记录计算机计算耗费的时间如表3所示。从表3看出,当小波包分解层次为4时代价函数值为5592.46,计算机计算时间为0.109秒,代价函数和计算时间均相对较小,因此损伤识别小波包分解层次取4。

表3 Daubechies 20 不同分解层次的代价函数值和计算时间

分解层次 i	1	2	3	4
$S_l(E_j)(p=1.5)$	16621.50	11913.00	7728.20	5592.46
计算时间/s	0.032	0.047	0.062	0.109
分解层次 i	5	6	7	8
$S_l(E_j)(p=1.5)$	6554.22	6470.30	7904.85	11283.70
计算时间/s	0.172	0.343	0.671	1.373

注:计算机的 CPU 为 Intel(R)Core(TM)i5 M2430 2.40 GHz

3.3 损伤识别

3.3.1 自振频率

为了研究古木结构不同损伤程度的损伤识别指标,列出完好结构、损伤工况1、损伤工况2的自振频率 f_0、f_1、f_2 见表4所示。

表4 古木结构的自振频率

振型阶数	f_0/Hz	f_1/Hz	误差%	f_2/Hz	误差%
1	2.902	2.898	0.14	2.893	0.32
2	4.063	4.063	0.00	4.063	0.00
3	4.416	4.416	0.00	4.416	0.01
4	9.383	9.368	0.15	9.351	0.34
5	17.857	17.852	0.03	17.847	0.06
6	35.274	35.274	0.00	35.274	0.00
7	35.634	35.631	0.01	35.628	0.02

续表

振型阶数	f_0/Hz	f_1/Hz	误差%	f_2/Hz	误差%
8	36.069	36.059	0.03	36.048	0.06
9	38.985	38.901	0.22	38.798	0.48
10	40.936	40.678	0.63	40.366	1.39

注:误差 = $\dfrac{|f_i - f_0|}{f_0}(i=1,2)$

从表4可以看出古木结构的损伤对自振频率的影响非常小,损伤工况1与完好结构的自振频率最大误差仅为0.63%,损伤工况2与完好结构的自振频率最大误差仅为1.39%,看来利用结构自振频率的变化来发现古木结构的损伤十分困难,因此本文提出了基于小波包能量曲率差的损伤识别指标。

3.3.2 小波包能量曲率差

(1)损伤工况1的小波包能量曲率差

对古木结构损伤工况1损伤前后梁上27到35节点的竖向加速度响应进行小波包分解,选择小波函数Db20,分解层数为4层,得到16个小波包系数和能量值,分析前8个能量的小波包能量曲率差如图4所示。

从图4(b)、(f)、(i)可以看出,损伤工况1的损伤发生在30到32节点之间的损伤单元52、53位置,这正好是梁假定的损伤单元52、53所在位置,与损伤工况1假定的损伤位置完全吻合,可以判定在此位置发生了损伤,说明小波包能量曲率变化量可以用于古木结构的损伤定位。

由于篇幅有限,损伤工况2损伤前后的小波包能量曲率差就不一一列举了。

(2)损伤工况1、2的小波包能量曲率差(前8个分量叠加)

从图4看出损伤工况1的小波包能量曲率差的前8个能量图中只有部分图形能确定结构的损伤位置,在此基础上将损伤工况1、2的小波包能量曲率差前8个分量进行叠加来分析损伤定位的效果如图5所示。

由图5(a)、(b)看出,将损伤工况1、2损伤前后的小波包能量曲率差的前8个分量进行叠加,在节点31处发生突变,其值最大,损伤位置最明显,随着节点离跨中越远,损伤指标值越小,到梁两端的27和35节点时,损伤指标值达到最小,说明该指标对损伤位置较敏感,小波包能量曲率差的前8个分量叠加比小波包能量曲率差更能识别古木结构的损伤位置,它可以作为损伤识别指标进行古木结构的准确定位,并且损伤程度越大,该指标值越大。

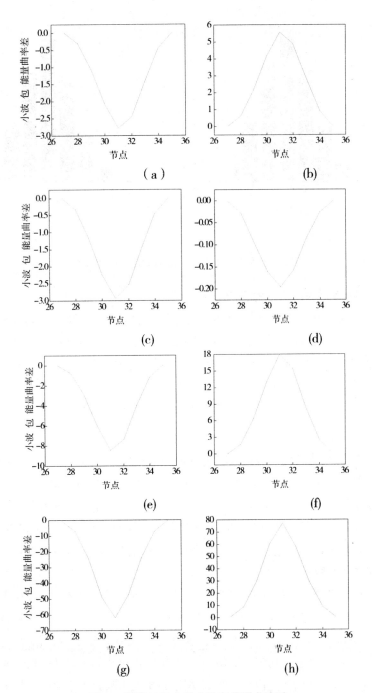

图4 损伤工况1的小波包能量曲率差

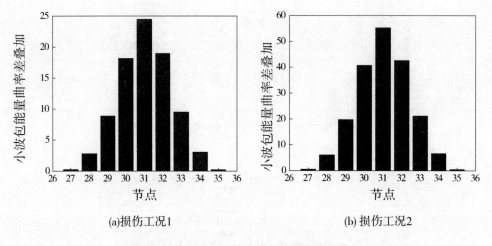

(a)损伤工况1　　　　　　　　　　　(b)损伤工况2

图 5　损伤工况 1、2 的小波包能量曲率差叠加

3.4　噪声对损伤识别的影响

在结构的健康监测过程中,由传感器采集的数字信号难免要受到外界噪声的干扰,对于信号中存在的噪声,一般假定为高斯白噪声,在原有信号基础上叠加服从正态分布均值为 0 的量作为测量噪声,通过信号的信噪比(SNR)来衡量信号的噪声水平,信噪比定义为:

$$SNR = 10\log \left| \frac{\sum x^2(n)}{\sum y^2(n)} \right| = 20\log \frac{A_S}{A_N} \text{(db)} \tag{23}$$

式中:A_S 为信号 x(n)的均方根,A_N 为噪声 y(n)的均方根。

下面以损伤工况 1 为例来分析在信噪比 SNR = 10、20、30、40、50 情况下对损伤定位效果的影响。在有限元分析得到的梁上各节点的竖向加速度信号中分别加入不同分贝的高斯白噪声来研究噪声对损伤识别的影响。仍选用 db20 小波函数进行小波包分解,分解层数为 4,不同噪声水平下的小波包能量曲率差的前 8 个分量叠加如图 6 所示。

从图 6 看出测试数据含有白噪声对信号的高频部分影响较大,故在含噪信号的小波包分解中应选用低频概貌信号和低频细节信号作为损伤识别的判别依据。当信噪比 SNR 小于或者等于 20db 时,该损伤指标受噪声影响较大,对损伤定位效果影响较大,已不具备损伤定位的能力。当信噪比 SNR 等于 30db 时,受噪声影响较小,已能基本进行损伤定位了。当信噪比 SNR 大于或者等于 40db 时,损伤定位能力已不受噪声影响,该损伤指标对损伤定位效果与无噪声信号基本相当,说明随着信噪比的提高,该损伤指标对损伤识别的敏感性逐渐增加,损伤定位效果越好,该损伤指标具有一定的抗噪声干扰能力。当信噪比较小时,受噪声的影响较大,因此需对含噪声信号进行消噪处理,尽可能还原为原始信号,才能保留损伤信息,以便对古木结构的损伤进行准确定位。

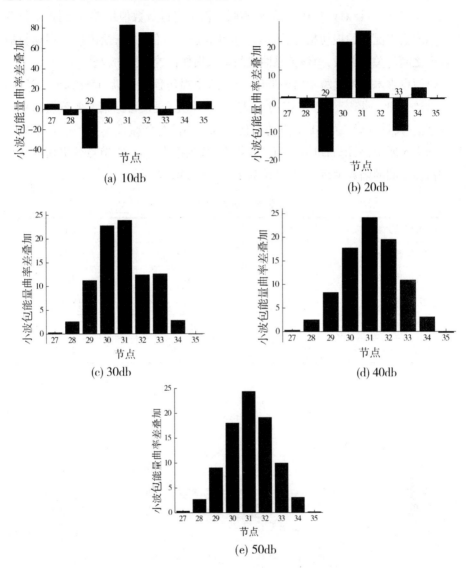

图6 不同噪声水平下小波包能量曲率差叠加

3.5 损伤程度的判定

由图5看出,对于无噪声信号在同一损伤位置不同损伤程度时,损伤指标柱状图基本相似,只是数值大小有差别。因此设想对于同一损伤位置不同损伤程度,若能找到损伤程度和损伤指标之间的函数关系,绘出其关系曲线,就能由该关

225

系曲线对损伤程度进行判断了。

损伤程度的判定方法:首先判定结构是否存在损伤,若存在损伤,再确定损伤的具体位置,然后针对该损伤位置,对不同损伤工况进行数值模拟,得到损伤位置上不同损伤工况下的损伤指标,再运用 matlab 进行数值拟合,绘出损伤指标与损伤程度之间的关系曲线,由该关系曲线判定该损伤位置的损伤程度。

仍假设古木结构梁跨中出现损伤,对梁跨中损伤程度 10%、15%、20%、25%、30%、35%、40%、45%、50% 各损伤工况进行有限元分析,得到梁上各节点的竖向加速度信号,计算梁跨中 31 节点的损伤指标—小波包能量曲率差的前 8 个分量叠加如表 5 所示,采用 matlab 进行数值拟合,找到损伤位置 31 节点处损伤程度和损伤指标之间的函数关系,绘制其函数关系曲线如图 7 所示。

表 5 损伤指标

损伤程度/%	10	15	20
损伤指标	24.49	38.98	55.29
损伤程度/%	25	30	35
损伤指标	73.78	94.79	118.83
损伤程度/%	40	45	50
损伤指标	146.43	178.24	214.92

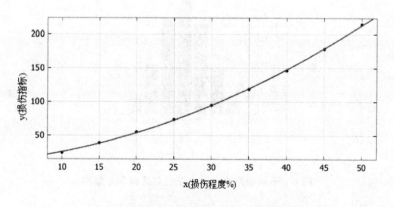

图 7 损伤程度和损伤指标之间的关系曲线

从图 7 得出损伤指标和损伤程度之间的函数关系式为:$y = 0.06236 x^2 + 0.9479 x + 10.13$,对古木结构损伤前后进行有限元模拟,得到梁上各节点的竖向加速度信号进行小波包分解,求出损伤指标,由该损伤指标就能在图 7 中找到相应的损伤程度了。

下面验算其适用性:假设梁跨中损伤程度为18%,对损伤工况3进行数值分析,得到梁上各节点的竖向加速度响应信号进行小波包分解,计算并绘制损伤指标如图8所示,得到31节点的损伤指标 y = 48.53,由图7可以逆推出 x = 18.35%,其相对误差为1.9%,可见其误差非常小,因此该函数关系式能对古木结构的损伤程度进行较准确的识别。

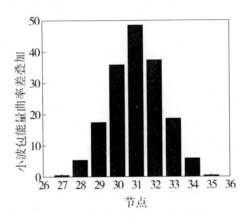

图8　损伤工况3的损伤指标

4　结　语

本文以西安钟楼为工程依托,对随机激励作用下的古木结构梁上各节点的加速度响应信号进行小波包分解,提出了小波包能量曲率差损伤识别指标,通过此指标进行古木结构的损伤定位,得出结论:

(1)在无噪声干扰下,该损伤指标对于古木结构的损伤定位比较敏感,可以准确判定古木结构的损伤位置,并且损伤指标随损伤程度的加大而增大。

(2)该损伤指标在高斯白噪声干扰下,当信噪比 SNR 大于或者等于40db时,损伤识别能力已不受噪声影响,能对古木结构的损伤进行准确定位,说明该损伤指标具有一定的抗噪声干扰能力,在实际工程应用中能够取得较好的定位识别效果。

(3)得出了损伤指标和损伤程度之间的函数关系式,用其进行损伤程度的判断并验算其适用性,为研究环境激励下西安钟楼的损伤预警提供了理论依据。

参考文献

[1]Coifman R R,Wickerhauser M V. Entropy – based algorithms for best basis selection [J].

IEEE Trans. Inf. Theory,1992,38:713-718.

[2]Doebling S W,Farrar C R,Prime M B. A summary review of vibration-based damage identification methods[J]. Shock Vib. Dig,1998,30(2):91-105.

[3]丁幼亮,李爱群,邓扬.面向结构损伤预警的小波包能量谱识别参数[J].东南大学学报(自然科学版),2011,41(4):824-828.

Ding You-liang,Li Ai-qun,Deng Yang. Parameters for identification of wavelet packet energy spectrum for structural damage alarming[J]. Journal of Southeast University(Natural Science Edition),2011,41(4):824-828.

[4]刘涛,李爱群,丁幼亮等.基于小波包能量谱的结构损伤预警方法试验研究[J].振动与冲击,2009,28(4):4-9.

Liu Tao,Li Ai-qun,Ding You-liang,et al. Experimental study on structural damage alarming method based on wavelet packet energy spectrum[J]. Journal of vibration and shock,2009,28(4):4-9.

[5]邓扬,丁幼亮,李爱群.基于小波包分析的拉索损伤声发射信号特征提取[J].振动与冲击,2010,29(6):154-158.

Deng Yang,Ding You-liang,Li Ai-qun. Feature extraction of acoustic emission signals for cable damage based on wavelet packet analysis[J]. Journal of vibration and shock,2010,29(6):154-158.

[6]范颖芳,胡志强,周晶,等.基于小波包分析的肋拱桥结构损伤状态研究[J].工程力学,2008,25(5):182-188.

Fan Ying-fang,Hu Zhi-qiang,Zhou Jing,et al. Study on damage state of ribbed arch bridge using wavelet packet analysis [J]. Engineering Mechanics,2008,25(5):182-188.

[7]HAN Jian-Gang,REN Wei-Xin,SUN Zeng-Shou. Wavelet packet based damage identification of beam structures [J]. International Journal of Solids and Structures,2005,42:6610-6627.

[8]余竹,夏禾,Goicolea J M,等.基于小波包能量曲率差法的桥梁损伤识别试验研究[J].振动与冲击,2013,32(5):20-25.

Yu Zhu,Xia He,Goicolea J M,et al. Experimental study on bridge damage identification based on wavelet packet energy curvature difference method [J]. Journal of vibration and shock,2013,32(5):20-25.

[9]李爱群,丁幼亮.工程结构损伤预警理论及其应用[M].北京:科学出版社,2007.

[10]孟昭博.西安钟楼的交通振动响应分析及评估[D].西安:西安建筑科技大学,2009.

[11]孟昭博,袁俊,吴敏哲,等.古建筑高台基对地震反应的影响[J].西安建筑科技大学学报(自然科学版),2008,40(6):835-840.

Meng Zhao-bo,Yuan Jun,Wu Min-zhe,et al. The influence of high-station base of ancient building on its seismic responses [J]. J. Xi'an Univ. of Arch. & Tech(Natural Science Edition), 2008,40(6):835 – 840.

[12]赵均海,俞茂宏,高大峰等.中国古代木结构的弹塑性有限元分析[J].西安建筑科技大学学报,1999,31(2):131 – 133.

Zhao Jun – hai, Yu Mao – hong, Gao Da – feng, et al. FEM analysis on the elasto – plasticity of ancient wooden structure [J]. J. Xi'an Univ. of Arch. &Tech,1999,31(2):131 – 133.

[13]韩广森.城市轨道交通微幅振动对古建筑的影响[D].西安:西安建筑科技大学,2011.

注:本文曾发表在2014年《振动与冲击》第4期上

地面交通激励下西安钟楼木结构的损伤识别

王 鑫 孟昭博*

本文对西安钟楼环道行驶的地面车辆激励下钟楼木框架结构进行损伤识别,选取钟楼上部木框架结构在第一层外围的中跨梁进行分析,对钟楼木框架结构采用小波包能量曲率差来进行损伤定位研究。研究表明:该指标在没有噪声干扰情况下能敏感地识别出钟楼上部木框架结构的损伤,能准确定位出钟楼木框架结构梁上出现的损伤位置;损伤程度增加,指标值增大。在信噪比SNR大于或等于40db情况下,有高斯白噪声的影响时能准确定位钟楼上部木框架梁上出现的损伤位置,能抵抗一定的噪声干扰影响;进一步判别了钟楼木框架结构的损伤程度并进行了验算,为古木建筑结构在交通激励下的损伤识别奠定了理论基础。

0 引言

我国是文明古国,有着悠久的历史文化。目前位于东西南北大街交汇处的西安钟楼是我国现存钟楼中体形最大、保存最完整的一个,它被誉为西安市的标志性建筑,1996年11月被国务院公布为全国重点文物保护单位。自上世纪80年代以来,钟楼二层地板上的展品由于地面交通车辆的微幅振动发生移位,高台基的墙面出现开裂裂缝,木构架横梁处的裂缝逐渐扩展。钟楼环道内侧到高台基边缘的距离虽然被扩大到了26.7m,但地面车辆的振动对钟楼的影响只增不减,站在钟楼二层地板上能明显感觉到有振感存在。西安是西部地区发展的重心,城市规模在不断扩大,人口继续不断增加,地面交通车辆日渐增多。钟楼环道地面车辆的行驶振动使钟楼木结构发生疲劳破坏、榫卯出现松动现象、木结构产生变形、钟楼的使用寿命在缩短。如果钟楼遭受到破坏的话,是不可再生的宝贵资源,产生不能弥补的损失,因此结构健康监测和安全评估系统需要第一时间发现结构的损伤并进行损伤预警。

---

* 作者简介:王鑫(1971—),女,陕西西安人,天水师范学院副教授,博士,主要从事结构健康监测研究。

小波分析有放大和缩小及平移等功能,有优良的时-频局部化特点,但是缺点是高频部分的分辨率差[1]。小波变换被小波包分析进行扩充,对信号能更加精细分析,在小波分析中原来没细分的高频部分被进一步分解,特性是有任意时-频分辨率,因而广泛应用于土木工程的健康监测和损伤诊断领域。

20 世纪 80 年代以来,在小波包分析方面许多国内外学者开展了大量研究工作,取得了可喜的研究成果。丁幼亮等[2]提出了结构的小波包能量谱和损伤预警指标来表示结构的损伤状态。刘涛等[3]做了钢梁的损伤试验目的是验证在实际工程中小波包能量谱损伤预警的有效性。孙鹏[4]设计制作悬索桥的试验模型,并进行动态测试,利用 Morlet 小波变换方法对模态参数识别的适用性和有效性研究,提出小波理论识别模态参数的改进方法。肖书敏[5]建立桥梁结构的有限元模型,对其进行动力特性分析,采用小波包分析方法处理结构动态响应信号来构造结构损伤指标,结合损伤指标和人工神经网络方法进行桥梁结构的损伤定位研究。余竹等[6]模拟了沧州子牙河新桥替换下来梁的两种损伤工况,结构的损伤采用小波包能量曲率差来进行识别。刘习军[7]对梁式结构利用小波包变换方法将结构的响应分解到不同的模态,引入曲率公式定义新的对损伤位置更加敏感的小波能量指标。

本文采用小波包能量曲率差从损伤敏感特性和噪声的影响方面对钟楼上部木框架结构进行损伤定位识别研究。

1 采用不等间距曲率的求解方法[6]

通常采用变量的二阶差分求曲率,即通过斜率的变化率来计算。

采用不等间距曲率的求解如下:

$$K_i = y_i^* = \frac{\dfrac{y_{i+1} - y_i}{h_{i+1}} - \dfrac{y_i - y_{i-1}}{h_{i-1}}}{\dfrac{h_{i+1} + h_{i-1}}{2}} \tag{1}$$

式(1)中分子为节点左段和右段的斜率差值,分母为节点左段和右端斜率差的间距。如果节点为等间距的话,则 $h_{i-1} = h_{i+1}$,那么

$$K_i = y_i^* = \frac{\dfrac{y_{i+1} - y_i}{h} - \dfrac{y_i - y_{i-1}}{h}}{h} = \frac{y_{i+1} - 2y_i + y_{i-1}}{h^2} \tag{2}$$

式(2)表示二阶差分法来求等间距曲率的计算公式。把式(2)中的 y 用小波包能量谱来替换的话,它则表示小波包能量曲率。运用小波包能量曲率来对完好结构和损伤工况的节点进行插值运算,得到在结构损伤工况下的小波包能量曲率差的

计算公式如下:

$$\Delta K_i = K_i^u - K_i^d \tag{3}$$

式(3)中 k_i^u, k_i^d 各表示完好结构与损伤结构的小波包能量曲率值。

2 西安钟楼有限元模型的建立

钟楼木框架结构通过外围 12 根木柱 Z-1、中间 4 根木柱 Z-2、木梁 L-1、L-2 及 L-3 构建而成,梁与柱的截面尺寸见表1,木材和土体材料参数见表2:

表1 梁与柱的截面尺寸[8]

柱	Z-1	Z-2	—
截面直径(mm)	500	700	—
梁	L-1	L-2	L-3
截面尺寸 b(mm)×h(mm)	300×700	300×800	200×300

表2 木材和土体材料参数[8,9]

材料	厚度(m)	弹性模量(MPa)	密度(kg/m³)	泊松比
木材	-	8307	410	0.25
台基夯土	8.6	20.9	1870	0.347
地基土	8	11.2	1780	0.33

木梁、木柱采用三维弹性单元 Beam188 来模拟,大屋盖采用 Mass21 质量单元来模拟,采用 Combinl4 单元模拟梁柱榫卯的连接。梁柱榫卯节点处建立多个有限元重合的节点,弹簧单元仅在梁与柱连接处施加,柱与柱间不加弹簧单元。弹簧单元把柱端节点①与梁端节点②③④⑤连接起来,采用 6 个弹簧单元来模拟柱端节点①与梁端节点的连接,通过输入梁柱六个自由度方向的弹簧刚度完成梁柱榫卯的连接如图 1 所示,采用钟楼监测数据计算的榫卯节点刚度的近似值 1×10^{10} kN·m/rad[8]。

高台基尺寸为 35.5m × 35.5m × 8.6m,考虑地面车辆在钟楼环道行驶,钟楼地基建成圆柱形地基,内圆柱地基半径为 44.45m,外圆台地基内半

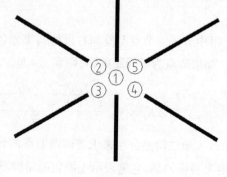

图1 榫卯节点

径为44.45m,外半径为65m,建模时把内圆柱地基和外圆台地基粘结在一起,使得钟楼环道网格划分较均匀、规则,地基深度取8m[10]。地基土和台基夯土采用Solid45单元,高台基和地基土的单元长度取2m,粘弹性人工边界采用Combin14单元,木柱与高台基相应节点耦合,体系的阻尼比$\xi_i = \xi_j = 0.03$[11]。为减小边界振动波的反射,土体的侧向边界采用粘弹性人工边界,土体的弹簧阻尼边界系数如表3所示[12]。

表3 土体边界参数表[12]

弹簧刚度系数			阻尼系数	
K_x (KN/m³)	K_y (KN/m³)	K_z (KN/m³)	c_p(压缩波) (KN·s/m)	c_s(压缩波) (KN·s/m)
16976	16976	16976	622	333

基底采用固定约束,地面为自由边,建立钟楼上部木结构、上部木结构—高台基、上部木结构—高台基—地基的有限元模型如图2~4所示。

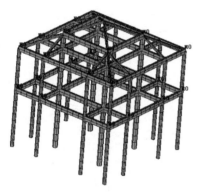 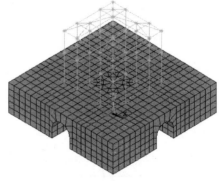

图2 木结构的有限元模型　　图3 木结构—高台基的有限元模型

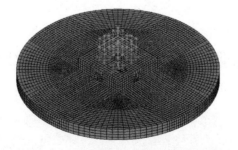

图4 木结构—高台基—地基的有限元模型

3 西安钟楼的损伤识别

采用最具代表的公交车来模拟地面交通激励对钟楼木框架结构的影响,研究地面车辆在钟楼环道以 20km/h 双线行驶对钟楼上部木框架的损伤识别。车辆荷载沿竖向施加在钟楼环道的单元节点上来模拟实际的车轮荷载作用,车辆荷载表达式为:$F(t) = 35000 + 2.16\sin(3t)$,车辆荷载的时程曲线和频谱图如图5所示[13],加载后的有限元模型如图6所示。

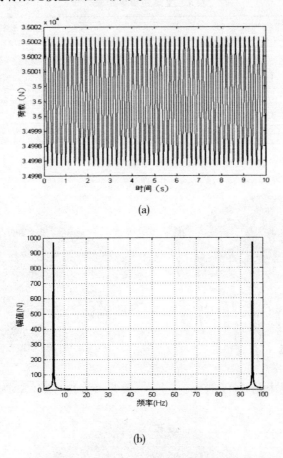

图 5　车辆荷载在 V = 20km/h 时的时程曲线(a)和频谱曲线(b)

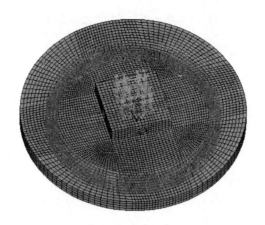

图 6　钟楼车辆双线行驶加载后的有限元模型

选择钟楼上部木框架结构跨度较大,损伤敏感性较强的第一层外围中跨梁进行损伤识别研究,损伤的具体位置及对应的节点详图如图 7 所示,钟楼木框架结构各损伤工况见表 4。损伤模拟时假定损伤构件的弹性模量 E 分别降低 $\triangle E = 10\%$、18%、20%[14],下面从损伤敏感性和噪声影响进行损伤识别研究。

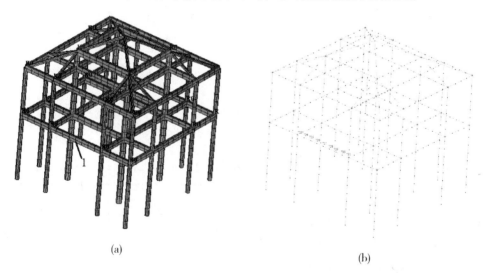

图 7　钟楼上部木框架梁的损伤位置(a)和对应的节点详图(b)

表4 钟楼木框架结构的损伤工况

损伤工况	损伤所在区域	损伤程度/%	备注
1	1(节点726、727、728)	10	一层外围中跨梁跨中附近
2	1(节点726、727、728)	20	一层外围中跨梁跨中附近
3	1(节点726、727、728)	18	一层外围中跨梁跨中附近

3.1 选择小波函数

对钟楼木框架完好结构在第一层外围中跨梁的跨中节点727处的竖向加速度采用Daubechies进行小波包分解,分解为4层,l^p范数熵代价函数值的计算见表5。小波阶次为7时,l^p范数熵最小,因而小波函数选定Daubechies7。

表5 分解层次取4不同小波阶数的代价函数值

小波阶数 N	5	6	7	8
$Sl(Ej)(p=1.5)$	7.20E-04	7.66E-04	7.19E-04	7.48E-04
小波阶数 N	9	10	11	12
$Sl(Ej)(p=1.5)$	7.77E-04	7.30E-04	7.62E-04	7.80E-04
小波阶数 N	13	14	15	16
$Sl(Ej)(p=1.5)$	7.37E-04	7.70E-04	7.81E-04	7.40E-04
小波阶数 N	17	18	19	20
$Sl(Ej)(p=1.5)$	7.76E-04	7.81E-04	7.45E-04	7.84E-04
小波阶数 N	21	22		
$Sl(Ej)(p=1.5)$	7.80E-04	7.47E-04		

3.2 选择小波包分解层数

对完好结构梁跨中节点727处的竖向加速度采用Daubechies7进行小波包分解,分解为1~8层,l^p范数熵代价函数值大小和计算所用时间长短见表6。由表6看出分解层次越多,代价函数值则越小。因而考虑计算所用时间长短,分解为6层。

表6　在不同分解层次采用Daubechies7的代价函数值和计算所用时间

分解层次 i	1	2	3	4
Sl(Ej)(p=1.5)	8.00E-04	8.00E-04	8.00E-04	7.19E-04
计算时间/s	0.031	0.031	0.048	0.093
分解层次 i	5	6	7	8
Sl(Ej)(p=1.5)	5.09E-04	4.09E-04	3.32E-04	1.78E-04
计算时间/s	0.177	0.296	0.609	1.212

3.3　损伤工况1和损伤工况2的小波包能量曲率差分析

对钟楼上部木框架结构在损伤工况1和损伤工况2下损伤前后梁上节点723到节点731间的竖向加速度采用Daubechies7进行小波包分解,分解为6层,得出64个小波包系数和小波包能量值,其中$f_6^1(t)$频带最低,$f_6^{64}(t)$频带最高。

在理论上讲,需要计算每一个信号分量对应的能量值大小,但是噪声的干扰对高频部分的影响非常大,因而对钟楼上部木框架结构只取前8个小波包分量进行损伤定位研究,绘制出前8个能量的小波包能量曲率差与梁上对应节点的关系图形如图8、9所示。

从图8(a)、(b)、(c)、(d)和图9(a)、(b)、(c)、(d)、(g)、(h)发现节点727处小波包能量曲率差突起特别明显,损伤工况1在节点726到节点728间的损伤单元288和289出现损伤,与所假设的损伤位置吻合,由此判断此处结构出现损伤,表明对钟楼木框架结构采用小波包能量曲率能进行损伤定位识别。

3.4　把小波包能量曲率差前8个分量在损伤工况1和损伤工况2情况下进行叠加

从图8、9发现损伤工况1和损伤工况2小波包能量曲率差前8个能量图只有几个图形能判别钟楼上部木框架结构的损伤位置,把小波包能量曲率差前8个分量在损伤工况1和损伤工况2情况下进行叠加如图10所示。

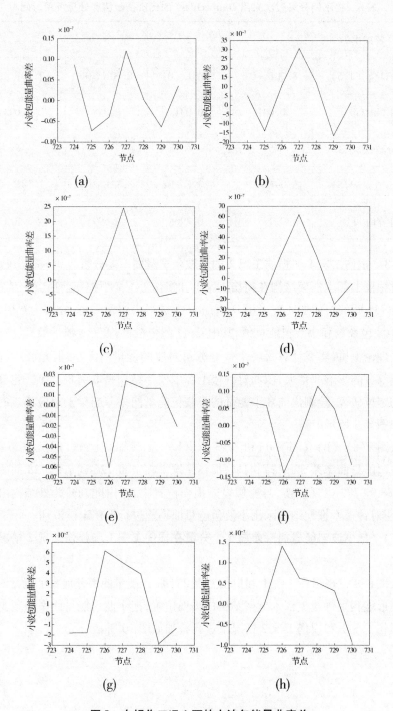

图 8　在损伤工况 1 下的小波包能量曲率差

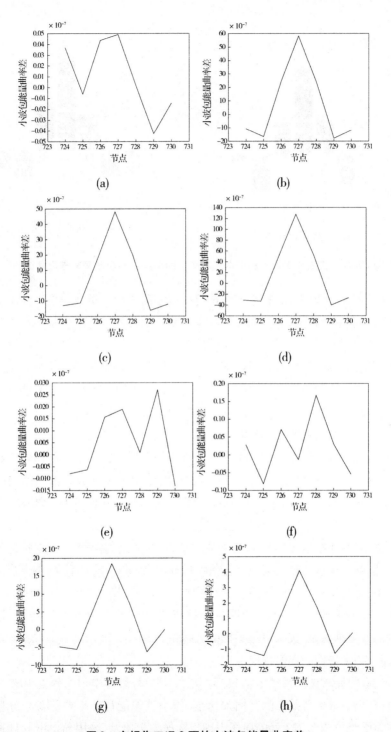

图 9 在损伤工况 2 下的小波包能量曲率差

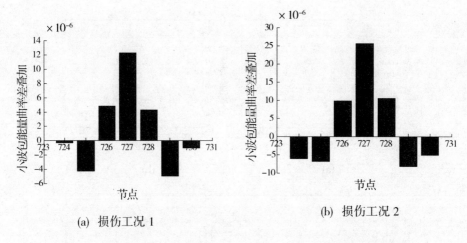

(a) 损伤工况 1　　　　　　　(b) 损伤工况 2

图 10　把损伤工况 1 和损伤工况 2 下的小波包能量曲率差进行叠加

从图 10(a)、(b)看出,把小波包能量曲率差前 8 个分量在损伤工况 1 和损伤工况 2 情况下进行叠加,节点 727 位置处突变特别明显,峰值出现跳变,可判断节点 727 附近的 288 和 289 单元出现损伤,与所假定的损伤位置吻合。离跨中越远的节点处损伤指标值越小,表明该指标对损伤定位识别很敏感。与小波包能量曲率差相比,把小波包能量曲率差前 8 个分量进行叠加更加能定位出钟楼木框架结构的损伤部位,该指标随着损伤程度的增加而增大。

3.5　噪声对识别效果的影响

研究在信噪比 SNR = 10 – 50 干扰下对损伤工况 1 损伤识别的影响。在钟楼木框架第一层外围中跨梁上节点的竖向加速度中加上不同分贝的高斯白噪声干扰,采用 Daubechies7 进行小波包分解,分解为 6 层,把小波包能量曲率差前 8 个分量在不同噪声干扰下进行叠加见图 11。

由图 11 发现在信噪比 SNR 小于或等于 20db 情况下,噪声干扰对指标值影响很大,不能对结构进行损伤定位。在信噪比 SNR 为 30db 情况下,噪声干扰对指标影响较小,可基本进行损伤定位。损伤定位能力在信噪比 SNR 大于或等于 40db 情况下已经不受噪声的干扰。表明信噪比越高,损伤定位的敏感性越强,损伤定位识别的效果越明显,抗噪声的干扰能力越强。

3.6　判定结构的损伤程度

采用有限元软件对钟楼木框架第一层外围中跨梁跨中损伤程度 10% ~ 50% 进行分析,提取梁上各节点的竖向加速度,把小波包能量曲率差前 8 个分量进行叠加作为梁跨中节点 727 处的损伤识别指标如表 7 所示,采用 matlab 软件通过数值拟合方法得到节点 727 处损伤程度与损伤识别指标的关系曲线如图 12 所示。

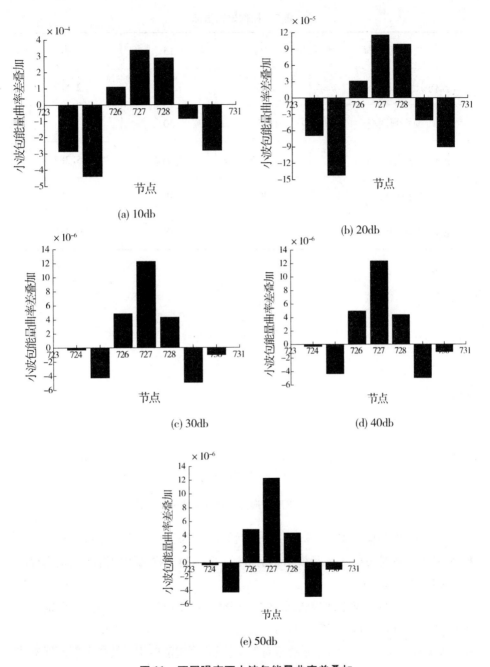

图 11 不同噪声下小波包能量曲率差叠加

表7 损伤识别指标

损伤程度/%	10	15	20
损伤识别指标($\times 10^{-5}$)	1.228	1.784	2.567
损伤程度/%	25	30	35
损伤识别指标($\times 10^{-5}$)	3.508	4.312	5.382
损伤程度/%	40	45	50
损伤识别指标($\times 10^{-5}$)	6.513	7.951	9.388

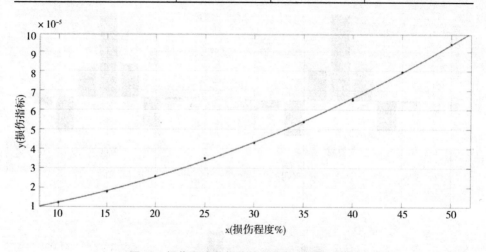

图12 损伤程度与损伤识别指标的关系曲线

从图12得到损伤识别指标与损伤程度的关系为:$y = 0.0024x^2 + 0.0592x + 0.4045$,其中 x 指损伤程度,y 指损伤识别指标。若求得了损伤识别指标 y,损伤程度 x 在图12就可找到了。

进一步验算:若梁出现损伤,跨中损伤18%,绘制损伤工况3的损伤识别指标如图13所示,得出节点727的损伤识别指标 $y = 2.553 \times 10^{-5}$,从图12逆推得出损伤程度 $x = 18.81\%$,仅有4.5%的相对误差,因此该式能应用于钟楼上部木框架结构损伤程度的准确判断。

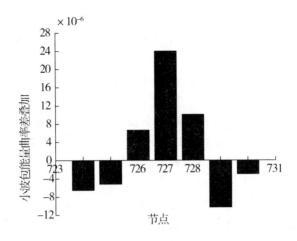

图13 损伤工况3的损伤识别指标

4 结 语

对西安钟楼环道行驶的地面车辆激励下钟楼木框架结构进行损伤识别研究,选取钟楼上部木框架结构在第一层外围的中跨梁进行分析,提出小波包能量曲率差对钟楼木框架结构进行损伤定位研究,得出以下结论:

(1)如果没有噪声的干扰,该指标能敏感地定位出钟楼上部木框架结构出现的损伤位置;损伤程度越大,指标值越大。

(2)在信噪比SNR大于或等于40db情况下,损伤识别已不受噪声的干扰,能够准确定位出钟楼上部木框架结构出现的损伤位置,能抵抗一定噪声干扰的影响。

(3)判别了钟楼木框架结构的损伤程度,并进行了验算,为古木建筑结构在交通激励下的损伤识别奠定了理论基础。

参考文献

[1]杨福生.小波变换的工程分析与应用[M].北京:科学出版社,2001.

[2]丁幼亮,李爱群,邓扬.面向结构损伤预警的小波包能量谱识别参数[J].东南大学学报(自然科学版),2011,41(4):824-828.

[3]刘涛,李爱群,丁幼亮等.基于小波包能量谱的结构损伤预警方法试验研究[J].振动与冲击,2009,28(4):4-9.

[4]孙鹏,丁幼亮,李爱群.利用Morlet小波变换识别悬索桥模型模态参数[J].振动、测试与诊断,2012,32(2):238-243.

[5]肖书敏,闫云聚,姜波澜.基于小波神经网络方法的桥梁结构损伤识别研究[J].应用

数学和力学,2016,37(2):149-159.

[6] 余竹,夏禾,Goicolea J M 等.基于小波包能量曲率差法的桥梁损伤识别试验研究[J].振动与冲击,2013,32(5):20-25.

[7] 刘习军,商开然,张素侠等.基于小波包变换的梁式结构损伤定位方法[J].实验力学,2015,30(3):305-312.

[8] 孟昭博.西安钟楼的交通振动响应分析及评估[D].西安:西安建筑科技大学,2009.

[9] 西安钟楼工程地质勘察报告[R].西安:机械工业勘察设计研究院,1985.

[10] 仇敏玉,俞亚南.道路行车荷载影响深度分析[J].岩土力学,2010,31(6):1822-1826.

[11] 袁晓铭,孙锐,孙静等.常规土类动剪切模量比和阻尼比试验研究[J].地震工程与工程振动,2000,20(4):133-139.

[12] 陈瑞春.西安地铁列车振动对钟楼影响的研究[D].北京:北京交通大学,2008.

[13] 梁志闯.交通随机荷载作用下西安城墙结构动力响应分析[D].西安:西安建筑科技大学,2013.

[14] 李爱群,丁幼亮.工程结构损伤预警理论及其应用[M].北京:科学出版社,2007.

注:本文曾发表在2017年《建筑结构》第9期上

西安钟楼的地震响应分析

王 鑫 孟昭博[*]

为了研究高台基和土-结构相互作用对结构地震反应的影响，本文以西安钟楼为研究对象，建立了钟楼的上部木结构、上部木结构—高台基、上部木结构—高台基—地基的三维有限元模型。选取了兰州波按照多遇地震和罕遇地震进行了地震反应分析，得出高台基对上部木结构的地震响应起放大作用，表明地震响应应该考虑高台基的影响。当考虑土与结构相互作用时，顶层的振动相对于模型1和模型2分别放大了13.53倍和2.27倍，因此对钟楼结构进行地震响应分析时必须考虑土与结构的相互作用。

引言

中国是文明古国，有着丰富的历史文化遗产，是中华民族乃至世界建筑艺术的瑰宝，其中古木结构最具中国特色。古建筑在几百年甚至上千年使用过程中，由于环境的侵蚀、材料老化和荷载的长期作用等使结构出现了损伤积累，致使抵抗自然灾害等能力下降。因此有必要了解古建筑木结构的特点及动力特性，为古建筑木结构的加固、维护和修缮提供理论依据。

西安钟楼是我国至今最宏伟、保存最完整的明代建筑之一，它处在东大街、西大街、南大街、北大街汇交处，是西安市的标志性建筑物[1]。钟楼是一个具有重檐三滴水、四角攒尖的古木建筑结构，建筑面积为1377.6m^2，它由台基和楼身主体及攒顶构成（见图1）。台基为正方形，边长为35.5m，高度为8.6m。外面采用青砖砌筑而成，里面回填夯土，在台基的四面正中央有高度和宽度各6m的券洞[2-4]。

---

[*] 作者简介：王鑫(1971—)，女，陕西西安人，天水师范学院副教授，博士，主要从事结构健康监测研究。

图1 西安钟楼

许多学者对西安钟楼上部木结构的抗震性能进行了研究,陈平[4]等建立二维梁-柱模型分析了西安钟楼的抗震性能,俞茂宏等[5]对西安钟楼进行了有限元分析,得出地震反应的主要模态,但是他们建立的模型都没有考虑高台基和木结构的共同作用问题。本文采用 ANSYS 有限元软件建立钟楼上部木结构(模型1)、上部木结构-高台基(模型2)、上部木结构-高台基-地基(模型3)的三维有限元模型,并采用兰州波对上述3种模型的动力特性和地震响应进行了分析。

1 西安钟楼的有限元模型和材料参数

钟楼木框架由外围12根木柱 Z-1、中间四根木柱 Z-2、木梁 L-1、木梁 L-2、木梁 L-3 构成,梁、柱截面尺寸见表1所示,材料参数见表2所示:

表1 梁、柱截面尺寸[6]

柱构件	Z-1	Z-2	—
截面直径(mm)	500	700	—
梁构件	L-1	L-2	L-3
截面尺寸 $b \times h$(mm^2)	300×700	300×800	200×300

表2 材料参数[6,7]

材料	弹性模量(MPa)	密度(kg/m^3)	泊松比
木材	8307	410	0.25
台基	20.9	1870	0.347
土体	11.2	1780	0.33

木梁、木柱采用三维弹性单元 Beam188 进行模拟,大屋盖采用 Mass21 质量单

元模拟,采用Solid45单元模拟高台基和地基,梁柱榫卯连接采用Combinl4单元来模拟。在进行有限元模拟时,在梁柱榫卯节点位置建立了多个有限元重合的节点,仅在梁和柱连接的位置施加弹簧单元,柱和柱之间不再施加弹簧单元。弹簧单元将柱端节点①和梁端节点②③④⑤分别连接了起来,柱端节点①和梁端节点采用了6个弹簧单元进行模拟,再输入梁柱六个自由度方向的弹簧刚度来实现梁柱榫卯的连接如图2所示,榫卯节点刚度值如表3所示[8]:

表3 榫卯节点刚度表

名称	$K_x = K_z(N/m)$	$K_y(N/m)$	$R_X = R_Y = R_Z(N \cdot m)$
边跨	1.71×10^7	2.07683×10^8	6.244×10^8

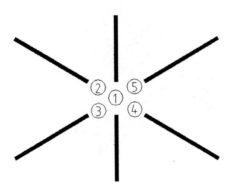

图2 榫卯节点

文献[9]取土体宽度为建筑物宽度的2倍,文献[10]取厚度为建筑物宽度的1/2来模拟半无限域土体,网格尺寸与土体厚度之比在1/3范围内较合理[9]。本文选取地基土尺寸为72m×72m×36m,土体网格大小取2m×2m×2m(见图5)。为减小截断边界带来的误差,采用了人工边界,土体的弹簧阻尼吸收边界系数[11]见表4所示。

表4 边界参数

弹簧刚度系数			阻尼系数	
$k_y(KN/m^3)$	$k_y(KN/m^3)$	$k_z(KN/m^3)$	c_p(压缩波)(KN·s/m)	c_s(压缩波)(KN·s/m)
16976	16976	16976	622	333

本文建立钟楼上部木结构(模型1)、上部木结构—高台基(模型2)、上部木结构—高台基—地基(模型3)的有限元模型如图3~5所示。

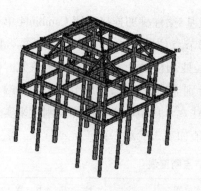
图3 木结构(模型1)

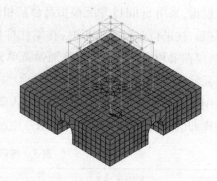
图4 木结构—高台基(模型2)

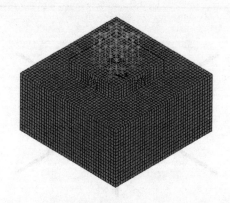
图5 木结构—高台基—土体结构(模型3)

2 钟楼的动力特性分析及结果

本文采用 Block Lanczos 方法,指定提取模态阶数为10阶,扩展模态也为10阶,模态分析计算结果如表5所示:

表5 自振频率

阶数		1阶	2阶	3阶	4阶	5阶	6阶	7阶	8阶	9阶	10阶
频率(Hz)	模型1	1.0880	1.3267	2.2580	3.6167	3.7845	4.5502	4.9598	5.0108	6.0767	6.1871
	模型2	0.6210	0.7287	1.1926	3.5303	3.6847	3.7708	3.7712	4.1246	4.1568	4.4795
	模型3	0.6117	0.7276	0.7368	0.7662	0.7688	0.9449	0.9625	0.9991	1.1029	1.1263

本文只列出了模型3的前六阶振型(见图6),从振型图来看,第一、二阶振型

表现为上部木结构沿 X 方向的平面滑移振动,第三、四阶振型为上部木结构与高台基整体的竖向振动,第五、六阶振型主要表现为弯曲和扭转振动,主要表现在梁柱的弯曲和扭转,高台基也发生了一定的变形。钟楼整体结构的低阶振型以木结构振动为主,第三阶后的振型为木结构和高台基同时振动,尤其高台基扭转时会更明显些,因此研究高台基古木结构的动力特性应该考虑高台基的影响作用。

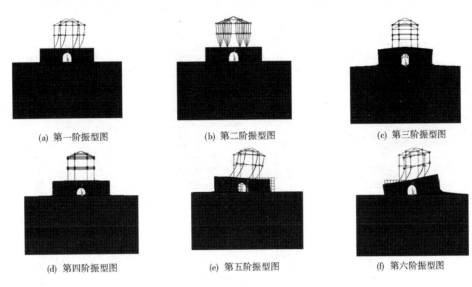

图6　模型3前六阶振型图

3　钟楼的地震响应分析及结果

本文采用 Rayleigh 定义的粘性比例阻尼,在结构抗震分析中取 $\xi_i = \xi_j = 0.05$ [12]。采用时程分析法进行钟楼的地震反应分析,考虑到西安市抗震设防烈度为 8 度,设计基本地震加速度值为 0.20g,选用兰州波,调整其幅值为 70gal(多遇地震)和 400gal(罕遇地震),分别输入 3 个模型进行计算,对其计算结果进行分析。以各层柱顶为研究对象,得到其加速度与位移时程曲线,部分位移与加速度时程曲线如图 7 所示:

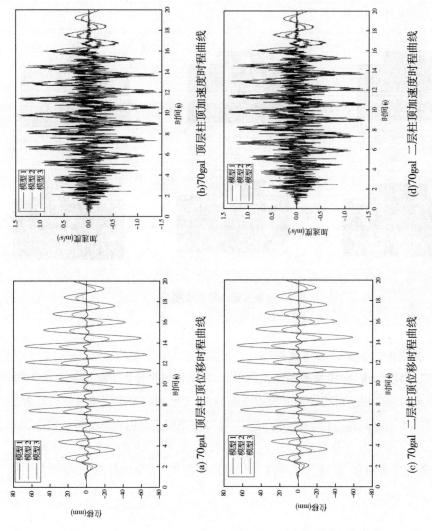

图7 兰州波作用下三个模型的时程曲线

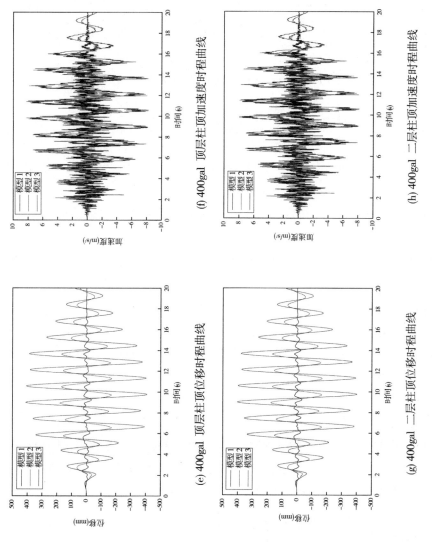

图 7 兰州波作用下三个模型的时程曲线（续）

各层结构的最大位移和加速度如表6所示。

表6 各层的最大位移和最大加速度

		位移(mm)				加速度(m/s^2)			
		台基顶	一层柱顶	二层柱顶	三层柱顶	台基顶	一层柱顶	二层柱顶	三层柱顶
70gal	模型1	—	3.36	5.12	5.31	—	0.70	0.78	0.79
	模型2	0.79	26.49	31.12	31.59	0.755	0.839	0.903	0.917
	模型3	14.95	59.95	70.59	71.85	0.99	1.24	1.40	1.45
400gal	模型1	—	19.22	29.23	30.31	—	4.00	4.46	4.49
	模型2	4.52	151.38	177.85	180.49	4.31	4.80	5.16	5.24
	模型3	85.43	342.54	403.39	410.59	5.66	7.11	8.01	8.26

由表6看出,在兰州波的作用下模型3考虑了土与结构的相互作用,其响应最大。在多遇地震作用下,顶层的位移和加速度的最大值分别达到71.85mm、1.45m/s^2,顶层柱顶的位移是模型2的2.27倍,是模型1的13.53倍;其次模型2的三层柱顶的位移和加速度的最大值分别为31.59mm、0.917 m/s^2,二层柱顶的位移是模型1的5.95倍。罕遇地震作用下,模型3的顶层的位移和加速度的最大值分别达到410.59mm、8.26m/s^2,顶层柱顶的位移是模型2的2.27倍,是模型1的13.55倍;其次模型2的顶层柱顶的位移和加速度的最大值分别为180.49mm、5.24 m/s^2,二层柱顶的位移是模型1的5.95倍。由于钟楼一、二层层高较大,三层层高较小,因而在地震作用下模型3的顶层柱顶加速度最大值稍大于二层柱顶。

由于高台基比上部木结构的刚度大,对木结构的地震响应起了放大作用,因此地震响应分析必须考虑高台基的影响。地基的刚度较小,考虑土与结构相互作用时,顶层柱顶的最大位移远大于模型1和模型2,因此对钟楼结构进行地震响应分析时必须考虑土与结构相互作用。

4 结论

本文建立钟楼的上部木结构、上部木结构－高台基、上部木结构－高台基－地基的三维有限元模型,对其进行动力特性和地震响应分析,得出以下结论:

(1)对模型1、2、3进行模态分析,得到其前10阶自振频率和振型,三种模型的基频分别是1.0880Hz、0.6210Hz、0.6117Hz。

(2)高台基对木结构起了放大作用,地震响应分析应该考虑高台基的影响。

(3)采用兰州波进行地震响应分析,3个模型的最大位移均出现在顶层柱顶。在多遇地震作用下,模型3的顶层柱顶的位移是模型2的2.27倍,是模型1的13.53倍;在罕遇地震作用下,模型3的顶层柱顶的位移是模型2的2.27倍,是模型1的13.55倍,因此对钟楼结构进行地震响应分析必须考虑土与结构相互作用。

参考文献

[1]陕西省文物局.陕西文物古迹大观[M].西安:三秦出版社,2003.

Shaanxi provincial administration of cultural heritage. Shaanxi cultural relics observatory[M]. Xi'an:Integrating Ecology Press,2003. (in Chinese)

[2]赵立瀛.陕西历史建筑[M].西安:陕西人民出版社,1992.

ZHAO Liying. Shaanxi historic buildings[M]. Xi'an:Shaanxi People's Press,1992. (in Chinese)

[3]杨绍武.西安钟楼[J].陕西档案,2010,(6):54-55.

YANG Shaowu. Xi'an bell tower[J]. Shaanxi Archive,2010,(6):54-55. (in Chinese)

[4]陈平,姚谦峰,赵冬.西安钟楼抗震能力分析[J].西安建筑科技大学学报,1998,30(3):277-280.

CHEN Ping, YAO Qianfeng, ZHAO Dong. An analysis on the aseismic behavior of Xi'an bell tower[J]. J. Xi'an Univ. of Arch. & Tech.,1998,30(3):277-280. (in Chinese)

[5]俞茂宏,张学彬,方东平.西安古城墙研究-建筑结构和抗震[M].西安:西安交通大学出版社,1993.

YU Maohong, ZHANG Xuebin, FANG Dongping. Research on the city wall in Xi'an structure and earthquake proof[M]. Xi'an:Xi'an Jiaotong University Press,1993. (in Chinese)

[6]孟昭博.西安钟楼的交通振动响应分析及评估[D].西安:西安建筑科技大学,2009.

MENG Zhaobo. Analysis and assessment of the vibration responds traffic-induced of Xi'an bell tower[D]. Xi'an:Xi'an University of Architecture and Technology,2009. (in Chinese)

[7]西安钟楼工程地质勘察报告[R].西安:机械工业勘察设计研究院,1985.

Xi'an bell tower engineering geological investigation report[R]. Xi'an:Survey and Design Institute of Mechanical Industry,1985. (in Chinese)

[8]韩广森.城市轨道交通微幅振动对古建筑的影响[D].西安:西安建筑科技大学,2011.

HAN Guangsen. The micro vibration of urban rail transit impact on ancient building[D]. Xi'an:Xi'an University of Architecture and Technology,2011. (in Chinese)

[9]吴体,熊峰,王永维.土-结构相互作用体系自振频率计算模型[J].四川建筑科学研

究,2006,32(4):95-99.

WUTi, XIONG Feng, WANG Yongwei. Research of model about soil – superstructure interaction system[J]. Sichuan Building Science,2006,32(4):95-99.(in Chinese)

[10]邓晓红,申爱国.土-结构相互作用中人工边界的影响[J].四川建筑,2006,26(6):86-89.

DENG Xiaohong, SHEN Aiguo. Influence of artificial lateral boundary in soil – structure interaction[J]. Sichuan Architecture,2006,26(6):86-89.(in Chinese)

[11]陈瑞春.西安地铁列车振动对钟楼影响的研究[D].北京:北京交通大学,2008.

CHENRuichun. Study on effects on bell tower due to train – induced vibrations on metro in Xi'an[D]. Beijing:Beijing Jiaotong University,2008.(in Chinese)

[12]袁晓铭,孙锐,孙静,孟上九,石兆吉.常规土类动剪切模量比和阻尼比试验研究[J].地震工程与工程振动,2000,20(4):133-139.

YUAN Xiaoming, SUN Rui, SUN Jing, MENG Shangjiu, SHI Zhaoji. Laboratory experimental study on dynamic shear modulus ratio and damping ratio of soil[J]. Journal of Earthquake Engineering and Engineering Vibration,2000,20(4):133-139.(in Chinese)

注:本文曾发表在2017年《世界地震工程》第9期上

一种基于脉冲耦合神经网络的图像双边滤波算法

刘 勍*

为有效滤除图像中高斯噪声干扰,提出了一种基于改进型脉冲耦合神经网络(PCNN)的双边滤波算法。首先从高斯噪声的特点出发,在改进PCNN中引入平滑抑制因子和自适应链接强度,并与相似神经元同步激活特性相结合,然后对含噪图像进行PCNN预滤波迭代处理,在滤除极值噪声的同时形成反映图像空时信息的时间矩阵,最后将时间矩阵信息运用在双边滤波中,并对其进行了自适应改进与滤波应用。实验结果表明,该算法在较好保护图像边缘细节等信息的情况下,能有效地滤除平滑区域噪声,去噪后图像具有较好的主观视觉效果和客观评价指标,信噪比高,去噪能力强,适应性好。

1 引言

图像信号在采集或传送过程中会受不同程度脉冲噪声、高斯噪声或混合噪声的干扰。而抑制或滤除图像噪声是图像处理领域研究的重要内容,一个好的滤波方法是在尽可能滤除噪声的同时能有效保护图像边缘和细节等信息。目前,针对图像高斯噪声的滤除,学者们提出了中值滤波、均值滤波等传统空域滤波方法以及基于小波域滤波[1]、非线性高阶统计量去噪[2]、偏微分方程降噪[3]及其他一些新型的滤噪算法[4]。双边滤波(bilateral filtering, BF)[5-8]是一种非线性空域滤波方法,近年来在图像高斯噪声滤波实践中倍受广大学者的广泛关注。

传统双边滤波在去噪过程中同时考虑了污染图像中邻域像素与中心像素间的几何距离和灰度相似性信息,并采用高斯核函数的形式进行表示,是一种非迭代加权滤波算法,在图像高斯噪声的滤除中取得了一定效果。但传统双边滤波中灰度部分只是简单利用了污染图像灰度相似信息,且选取固定方差[6],由于图像

---

* 作者简介:刘勍(1970—),男,甘肃天水人,天水师范学院教授、博士,主要从事图像信号处理及人工神经网络研究。

噪声与邻域像素灰度的独立性以及不同噪声污染图像统计信息的可变性,将致使其滤波适应性能较差;文献[7]等虽然提出了灰度高斯核函数方差随污染图像的全局方差线性变化关系,增强了双边滤波的自适应能力,但是没有充分体现局部方差的滤波特征,滤波效果也不太理想。

为此,本文提出了基于脉冲耦合神经网络[9](Pulse Coupled Neural Networks, PCNN)的图像双边滤波算法,首先在对 PCNN 模型链接强度改进及引进平滑抑制因子的基础上,利用相似神经元同步时空特性来迭代处理含噪图像,形成 PCNN 时间矩阵,然后将 PCNN 时间矩阵自适应地应用到双边滤波中,并用时间矩阵中神经元激活时刻的相似性替代污染图像的灰度相似性,最后通过实验验证了所提算法的有效性。

2 改进 PCNN 模型

传统 PCNN 是在哺乳动物视觉皮层模型启发下形成的第三代新型人工神经网络,构成 PCNN 的每一神经元由接受、调制和脉冲产生三部分组成,该模型已被广泛地应用于图像分割、边缘检测及目标识别等不同图像处理领域[10~13],显示了其优越性,但在具体不同应用中也存在一定缺陷。为克服传统 PCNN 模型神经元信号各部分间耦合的繁冗性、神经元点火状态的复杂性和处理时间的加长化,以及针对图像高斯噪声滤除中噪声方差变化造成的影响等,本文在文献[13,14]PCNN 模型的基础上对其进行了简化与改进,并用离散数学方程描述如下:

$$F_{ij}[n] = I_{ij}[n-1] \tag{1}$$

$$L_{ij}[n] = V_L \sum_{kl} W_{ijkl} Y_{ij}[n-1] - d \tag{2}$$

$$U_{ij}[n] = F_{ij}[n](1 + \beta_{ij}[n]L_{ij}[n]) \tag{3}$$

$$Y_{ij}[n] = \begin{cases} 1, & U_{ij}[n] > \theta_{ij}[n] \\ 0 & otherwise \end{cases} \tag{4}$$

$$\theta_{ij}[n] = \theta_0 \exp(-\alpha_\theta n) \tag{5}$$

$$T_{ij}[n] = \begin{cases} n, & Y_{ij}[n] = 1 \\ T_{ij}[n-1], & otherwise \end{cases} \tag{6}$$

下标 ij 为神经元的标号,n 为迭代次数(时间),I_{ij}、F_{ij}、L_{ij}、U_{ij}、θ_{ij}、T_{ij} 分别为神经元的外部刺激(图像像素构成矩阵中第 ij 个像素的灰度值)、反馈输入、连接输入、内部活动项、动态阈值和神经元激活时间,W 为链接权矩阵,V_L、θ_0 为幅度常数,α_θ 为相应的衰减系数,Y_{ij} 是二值输出。

为适应高斯噪声图像平坦区域的滤波和边缘细节的保留,改进 PCNN 模型中引入了平滑抑制因子 d 和自适应链接强度 β_{ij}，其中平滑抑制因子 d 能起到非线性滤波的作用，并且随着其值的增加，对图像的平滑度越明显，而自适应链接强度 β_{ij} 表示如下：

$$\beta_{ij}[n] = \frac{1 - (p-1)\left(\frac{\nabla I_{ij}[n-1]}{K}\right)^p}{\left[1 + \left(\frac{\nabla I_{ij}[n-1]}{K}\right)^p\right]^2} \tag{7}$$

其中 ∇I 为图像梯度，K 为扩散常数，当 $p = 1$ 时，$\beta_{ij} \geq 0$，式（3）内部活动项总是产生耦合增大；但当 p 由 1~2 变化时 β_{ij} 会出现负值，特别的当 $p = 2$ 时（本文中选取 $p = 2$），

$$\begin{cases} \beta_{ij} \geq 0, & 0 \leq \nabla I_{ij} \leq K \\ \beta_{ij} < 0, & \nabla I_{ij} > K \end{cases} \tag{8}$$

则在图像平坦区域，即 $|\nabla I_{ij}| < K$ 的区域，式（3）产生耦合增大，可对同一区域神经元进行扩大捕获，但在边缘附近，梯度足够大以致 $|\nabla I_{ij}| > K$，使式（3）产生耦合减小，形成反向扩大捕获，这就意味着保护了细节，使边缘得到了锐化。

在用 PCNN 进行图像处理时，先将输入图像的 $M \times N$ 个像素分别与 PCNN 网络的 $M \times N$ 个神经元相对应，然后进行迭代处理。其阈值随时间指数衰减，当某一神经元的内部活动项大于阈值时点火激活，被激活的神经元则通过与之相邻神经元的连接与平滑抑制关系激励邻接神经元，若邻接神经元的内部活动项大于阈值，则被捕获激活，其输出为 1，此时可将相似神经元捕获激活的时刻忠实记录在反映处理图像空时信息关系的对应时间矩阵中。因此，利用连接输入平滑抑制因子和自适应链接强度的调节，可以有效地触发处理区域内相似神经元的集体激活以及自动保护边缘神经元的状态信息，从而实现图像滤波去噪等处理。

3 PCNN 双边滤波算法

3.1 双边滤波及其改进

双边滤波是由 Tomasi 等人提出的一种非迭代、非线性局部滤波方法[6]。该方法在保护边缘细节的同时能够滤除噪声，其特点是在处理相邻各像素值时，不仅考虑到距离上的邻近关系，同时也考虑灰度上的相似性，通过对二者的非线性调制组合，可对信号进行滤波，特别适合于类高斯噪声的处理。其滤波模型如下所示：

$$\hat{I}_{i_0 j_0} = \sum_i \sum_j h(i_0 j_0, ij) \cdot I_{ij} \tag{9}$$

$$h(i_0j_0,ij) = \begin{cases} \dfrac{1}{C_{i_0j_0}}\exp\left(-\dfrac{(i-i_0)^2+(j-j_0)^2}{2\sigma_D^2}\right)\exp\left(-\dfrac{(I_{ij}-I_{i_0j_0})^2}{2\sigma_I^2}\right) & ij \in \Omega_{i_0j_0} \\ 0, & otherwise \end{cases}$$

(10)

$$C_{i_0j_0} = \sum_{i=i_0-\eta}^{i_0+\eta}\sum_{j=j_0-\eta}^{j_0+\eta}\exp\left(-\frac{(i-i_0)^2+(j-j_0)^2}{2\sigma_D^2}\right)\exp\left(-\frac{(I_{ij}-I_{i_0j_0})^2}{2\sigma_I^2}\right) \quad (11)$$

其中 I_{ij} 是噪声污染图像，$\hat{I}_{i_0j_0}$ 为滤噪后图像，$h(i_0j_0,ij)$ 是加权滤波因子，$\Omega_{i_0j_0}$ 为选择滤波窗口，i_0j_0 为窗口中心，$\Omega_{i_0j_0} = \{ij:ij \in [i_0-\eta,i_0+\eta] \times [j_0-\eta,j_0+\eta]\}$，$C_{i_0j_0}$ 是归一化权值因子，σ_D 是空域滤波时高斯函数的标准差，σ_I 是邻域中像素通过高斯函数进行滤波时的灰度标准差。

双边滤波器中利用高斯函数来进行空域和像素相似性的联合加权滤波系数调节，可以有效地滤除图像高斯噪声，其中像素相似性滤波系数选择是最为关键的因素。现有双边滤波方法是直接选择污染图像作为像素相似性度量的，其中 σ_I 选取固定值[5,8]或者根据图像输入信噪比按照经验公式计算[6]。但同一灰度相似区域的像素由于噪声污染造成其值各不相同，直接利用含噪图像作为双边滤波器加权系数的来源，将会影响到滤波效果，并且不同的随机噪声在图像中也会产生较大奇异值使得到的加权滤波系数不太合理，将会引起比较严重的失真结果。另外在实际图像去噪中图像的污染程度会随其环境发生变化且其信噪比也是未知的，这样也会影响滤波系数的自动选取，为克服上述问题，考虑 PCNN 相似神经元同步点火特性，采用改进 PCNN 对含噪图像迭代预处理，然后形成预滤波处理的时间矩阵 T_{ij}，最后在双边滤波联合加权系数选取中自适应选择时间矩阵对应时间像素的相似性，从而取代原含噪图像像素相似性，其表达式如下：

$$h(i_0j_0,ij) = \begin{cases} \dfrac{1}{C_{i_0j_0}}\exp\left(-\dfrac{(i-i_0)^2+(j-j_0)^2}{2\sigma_D^2}\right)\exp\left(-\dfrac{(T_{ij}-T_{i_0j_0})^2}{2\sigma_T^2}\right), & ij \in \Omega_{i_0j_0} \\ 0, & otherwise \end{cases}$$

(12)

$$C_{i_0j_0} = \sum_{i=i_0-\eta}^{i_0+\eta}\sum_{j=j_0-\eta}^{j_0+\eta}\exp\left(-\frac{(i-i_0)^2+(j-j_0)^2}{2\sigma_D^2}\right)\exp\left(-\frac{(T_{ij}-T_{i_0j_0})^2}{2\sigma_T^2}\right) \quad (13)$$

滤波模型与式(9)相同，σ_T 自适应选取 T 邻域中像素高斯函数的时间标准差。

3.2 滤波算法描述

从图像高斯噪声滤除的角度出发，在对 PCNN 模型简化改进的基础上，通过对含噪图像的预滤波处理，并将其迭代产生的时间矩阵应用在改进双边滤波中，

有效地达到对污染图像噪声进一步滤除的目的,具体滤波算法描述如下:

Step1:选取 V_L、θ_0、α_θ、d、K、η、σ_D 以及 W 等参数,确定总迭代时间(次数) n_0,当 $n=1$ 时,输入原始高斯噪声污染图像 $I[0]$,并按照式(7)计算 β_{ij};

Step2:按照改进型 PCNN 模型中式(1)~式(5)进行迭代处理;

Step3:在 $Y[n]$ 中以 $Y_{ij}[n]=1$ 像素为中心对其 3×3 邻域判断处理:如果邻域中除 $Y_{ij}[n]=1$ 外,其余值全为 0,则 $I[0]$ 中对应 ij 处像素用除 $I_{ij}[0]$ 外的邻域均值代替,否则,$I[0]$ 中对应 ij 处像素 $I_{ij}[0]$ 值保持不变,然后用 $I[n]$ 替代 $I[0]$;

Step4:按照式(6)形成 PCNN 时间矩阵 $T[n]$,如果 $T_{ij}[n]=0$,则令 $n=n+1$ 并转入 Step2,否则,转入 Step5;

Step5:按照式(12)计算双边滤波加权系数 $h(i_0 j_0,ij)$;

Step6:用预滤波处理图像 $I[n]$ 代替原始噪声污染图像,并按式(9)进行双边滤波。

Step7:输出去噪后的图像 $\hat{I}$。

4 实验结果与分析

为验证本文所提滤波算法的有效性,对受不同方差高斯噪声污染的多幅图像进行实验,并与文献[6]算法及文献[7]算法作了比较,实验中参数 $V_L=0.5$,$\theta_0=255$,$\alpha_\theta=0.09$,$d=0.3$、$K=100$、$\eta=2$、$\sigma_D=25$,W 为 3×3 元素全为 1 的矩阵,滤波性能用图像的峰值信噪比 PSNR 和均方误差 MSE 指标客观评价。

图 1(a1)~(c1),(a2)~(c2),……,(a6)~(c6)各列图像分别为 Lena、Bridge 和 Nature 原始图像,受方差为 0.003、0.005 和 0.007 高斯噪声污染图像,文献[6]及文献[7]滤波结果,PCNN 时间矩阵的灰度图像以及本文算法的滤波图像;图 2 及图 3 是对 Bridge 图像在受不同高斯噪声污染时,分别采用 3 种滤波算法的 PSNR 和 MSE(为作图比较方便,采用 ln(MSE))随噪声方差变化的关系曲线;表 1 为 3 幅图像在受方差为 0.004 高斯噪声污染时,分别采用 3 种滤波算法的 PSNR 和 MSE 对比实验数据。

对受高斯噪声污染的图像通过改进 PCNN 预处理,首先有效地滤除了散落在图像中的少量随机极值噪声,其次模型中由于采用了平滑抑制因子、自适应链接强度和相似神经元同步捕获激活等措施,由图 1(a5)~(c5)的时间矩阵图像可见,在较好保持原像图边缘等有用信息的前提下,图像平滑区域达到了初步滤除,为双边滤波联合加权系数的形成提供了良好的保障。另外,从上述图表不同的实

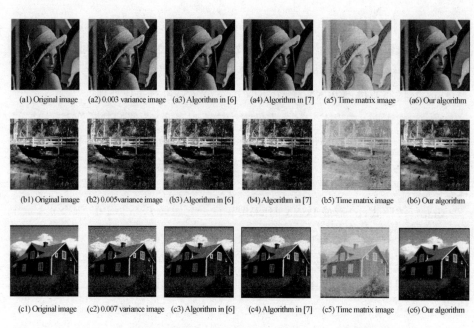

图 1 对 3 幅不同高斯噪声污染图像采用不同方法的滤波结果

验结果可以看出,对不同纹理和细节的图像在受不同噪声污染时,本文算法都具有良好细节信息保护和有效滤除平滑区域噪声的能力,去噪后图像主观视觉效果和客观评价指标 PSNR 以及 MSE 均好于文献[6]及文献[7]算法的滤噪结果,体现了通过对 PCNN 迭代预处理和非迭代双边滤波的良好互补结合,较好地克服了双边滤波非迭代性的一些不足;并且随着噪声方差的增加,本文算法视觉效果和客观评价指标在所给几种算法中一直保持良好的优势,进一步说明了本文算法在滤除高斯噪声污染图像上更加表现了良好的视觉效果、较强的去噪性能、稳定的鲁棒性和广泛的适应性。

表 1 对 3 幅噪声方差为 0.004 图像采用 3 种不同滤波算法的 PSNR 和 MSE 实验数据

Filter	Lena image		Bridge image		Nature image	
	PSNR	MSE	PSNR	MSE	PSNR	MSE
Algorithm in [6]	24.2441	244.7211	24.2088	246.7167	24.3542	238.5963
Algorithm in [7]	24.1307	251.1950	23.6973	277.5553	24.3239	241.6132
Our algorithm	25.8420	164.7282	25.0868	194.9747	25.5139	182.6777

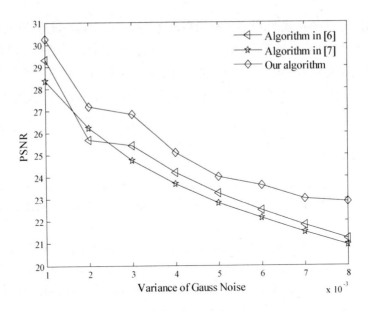

图 2　Bridge 图像去噪中采用不同算法时 PSNR 随噪声方差的变化关系

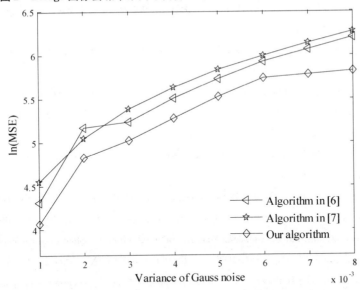

图 3　Bridge 图像去噪中采用不同算法时 ln(MSE) 随噪声方差的变化关系

5　结论

本文提出了一种基于改进型 PCNN 的双边滤波算法。在利用 PCNN 相似神经元同步激活特性的同时,在其链接部分和非线性耦合部分分别引入平滑抑制因子和采用自适应链接强度,并首先应用在含高斯噪声图像预滤波处理中,通过 PC-NN 迭代处理在滤除极值噪声的同时形成了反映图像空时信息的时间矩阵。然后将其与传统双边滤波器相结合,对双边联合滤波权值进行了自适应优化改进并应用在滤波实践中。最后通过实验仿真证明了所提算法是一种行之有效的滤除图像高斯噪声的去噪算法。

参考文献

[1] Yeqiu Li, Jianming Lu, Ling Wang, et al. Removal of Gaussian noise from degraded images in wavelet domain[J]. IEEJ Transactions on Electronics, Information and Systems, 2006, 126(11): 1351–1358.

[2] Hosny M. Ibrahim Reda R. Garieb Aboulnasr H. Ibrahim. Nonlinear Cumulant Based Adaptive Filter for Simultaneous Removal of Gaussian and Impulsive Noises in Images[C]. Proceedings of the Eighteenth National Radio Science Conference, Mansoura, Egypt, 2001, 269–76.

[3] Aubert G, Kormprobst P. Mathematical Problems in Image Processing: Partial Differential Equations and the Calculus of Varations[J]. Applied Mathematical Sciences, Springer – Verlag, 2001; 147

[4] Buades A, Coll B, and Morel J M. A review of image denoising algorithms, with a new one [J]. Multiscale Modeling Simulation, 2005, 4(2): 490–530.

[5] Ming Zhang, Bahadir Gunturk. A new image denoising method based on the bilateral filter [C]. IEEE International Conference on Acoustic, Speech and Signal Processes, Las Vegas, NV, USA, 2008, 929–932.

[6] C. Tomasi and R. Manduchi. Bilateral filtering for gray and color images[C]. in *Proc. Int. Conf. Comput. Vis.*, 1998, 839–846.

[7] Harold Phelippeau, Hugues Talbot, Mohamed Akil, et al. Shot Noise Adaptive Bilateral Filter[C]. 9th International Conference on Signal Processing, Beijing, China, 2008, 864–867.

[8] Buyue Zhang, Jan P. Allebach. Adaptive Bilateral Filter for Sharpness Enhancement and Noise Removal[J]. IEEE TRANSACTIONS ON IMAGE PROCESSING, 2008, 17(5): 664–678.

[9] R Eckhorn, H J Reitboeck, M Arndtetal. Feature linking via synchronization among distributed assemblies: simulation of results from cat cortex [J]. Neural Computation, 1990, 2(3): 293~307.

[10] H S Ranganath, G Kuntimad. Object detection using pulse coupled neural networks [J]. IEEE Trans, Neural Networks, 1999, 10(3): 615~620.

[11] Liu Qing, Ma Yi – de. Qian Zhi – Bai. Automated image segmentation using improved PC-NN model based on cross – entropy [J]. Journal of Image and Graphics, 2005, 10(5): 579~584.

刘勍,马义德,钱志柏.一种基于交叉熵的改进型 PCNN 图像自动分割新方法[J].中国图象图形学报.2005,10(5):579~584.

[12] Liu Qing, Ma Yi – de. A New algorithm for Noise Reducing of Image Based on PCNN Time Matrix[J]. Journal of Electronics & Information Technology,2008,30(8):1869 – 1873.

刘勍,马义德.一种基于 PCNN 赋时矩阵的图像去噪新算法[J].电子与信息学报,2008,30(8):1869 – 1873.

[13] Liu Qing, MA Yi – de, ZHANG Shao – gang, et al. Image Target Recognition Using Pulse Coupled Neural Networks Time Matrix[C]. Proceedings of the 26th Chinese Control Conference. Zhangjiajie, Hunan, China. 2007,7:96~99.

[14] Robert D. Stewart, Iris Fermin, Manfred Opper. Region Growing With Pulse – Coupled Neural Networks：

An Alternative to Seeded Region Growing[J]. IEEE TRANSACTIONS ON NEURAL NETWORKS,2002,13(6):1557~1562.

注：本文曾发表在 2012 年 10 月《International Journal of Computer & Applications》第 4 期上。

一种基于改进 PCNN 噪声检测的两级脉冲噪声滤波算法

刘勍[*]

为有效地滤除图像中严重脉冲噪声的干扰,提出了一种基于改进型脉冲耦合神经网络(PCNN)噪声检测的两级脉冲噪声去除算法。该算法首先利用 PCNN 同步脉冲发放特性区分定位噪声点和信号点位置,其次根据噪声点局部邻域信息对噪声点进行第一级自适应滤波,然后再利用具有保护边缘细节特点的多方向信息中值滤波器进行二级细微辅助滤波。实验结果表明,该算法在噪声检测中无需设定检测阈值,在有效滤除脉冲噪声的同时,能很好地保护图像边缘等细节信息,去噪后图像不但具有较好的主观视觉效果和客观评价指标,而且比传统中值滤波及其它相关算法具有更优的滤波性能,去噪能力强、信噪比高和适应性好,特别是对受严重噪声污染的图像,显示了更大的优越性。

1 引言

图像信号在生成和传输过程中会受到多种因素的干扰而感染噪声,从而导致图像质量下降,严重影响特征提取及模式识别等后续图像处理环节。脉冲噪声是一种典型的噪声类型,在滤除它的同时要求尽可能地保护图像边缘和细节等信息,为此,针对脉冲噪声的去除学者们提出了许多滤波方法[1-6],这些算法是传统中值滤波法(TM)或基于 TM 方法的改进,虽能达到一定的去噪效果,但会破坏和丢失相对于滤波窗口尺寸较小的图像细节,随着噪声的增加,滤波性能也会变得很差,且这些算法是对含噪图像中所有像素全部处理,这必然会损失图像中未污染像素信息;近几年,许多研究者又提出了一些新型的滤波算法[7-11],这些算法虽然先对含噪图像的像素进行检测分类,然后仅对噪声像素进行滤波处理,在去噪

---

[*] 作者简介:刘勍(1970—),男,甘肃天水人,天水师范学院教授、博士,主要从事图像信号处理及人工神经网络研究。

性能上有所提高,但它们对噪声点的定位依赖于阈值的选取,不利于自动噪点检测,自适应能力差,且在图像受到严重噪声污染时,去噪能力有限。

为此,本文提出了一种基于改进脉冲耦合神经网络(Pulse Coupled Neural Networks,PCNN)噪声检测的两级脉冲噪声滤波算法。该算法首先在利用 PCNN 相似神经元同步点火性质[12-15]自动检测噪声点的基础上,对噪声像素进行第一级自适应滤波以及对前一级滤波结果运用多方向信息中值滤波,最后通过实验证明了所提算法的有效性。

2 PCNN 改进模型

PCNN 也称为第三代人工神经网络,有一定的生物学背景,是 Eckhorn 等人依据猫、猴等动物大脑视觉皮层上的同步脉冲发放现象提出的,已被广泛地应用于图像平滑、分割、边缘检测及目标识别等图像处理领域[14-16],显示了其优越性。构成 PCNN 的每一神经元 N 由接收、调制和脉冲产生 3 部分组成,为克服传统 PCNN 人工设置参数多而且适应性能差的缺点,本文在文献[16]传统 PCNN 模型的基础上对其进行了简化与改进,其简化改进的离散数学方程描述如下:

$$F_{ij}[n] = I_{ij} \tag{1}$$

$$L_{ij}[n] = \begin{cases} 1, & \sum_{k,l \in W} Y_{kl}[n] > 0 \\ 0, & otherwise \end{cases} \tag{2}$$

$$U_{ij}[n] = F_{ij}[n](1 + \beta_{ij}[n]L_{ij}[n]) \tag{3}$$

$$Y_{ij}[n] = \begin{cases} 1, & (U_{ij}[n] \geq T_{ij}[n]) \\ 0, & (U_{ij}[n] < T_{ij}[n]) \end{cases} \tag{4}$$

$$\theta_{ij}[n] = \theta_0 e^{-\alpha_\theta(n-1)} \tag{5}$$

其中 ij 下标为神经元的标号,n 为迭代次数,I_{ij}、$F_{ij}[n]$、$L_{ij}[n]$、$U_{ij}[n]$、$\theta_{ij}[n]$ 分别为神经元的外部刺激(图像像素构成的矩阵中第 ij 个像素的灰度值)、第 ij 个神经元的第 n 次反馈输入、连接输入、内部活动项和动态阈值,W 为链接权矩阵,θ_0 为阈值幅度常数,自适应选取为图像像素灰度最大值 I_{max},α_θ 为相应的衰减系数,$Y_{ij}[n]$ 是 PCNN 的二值输出,β_{ij} 为链接系数,其值由下式决定:

$$\beta_{ij}[n] = \begin{cases} \max_{k,l \in W}\left(\dfrac{1}{(I_{ij} - I_{kl})^2 + 1}\right), & Y_{kl}[n] = 1 \\ 0, & otherwise \end{cases} \tag{6}$$

三个参数中 θ_0 和 β_{ij} 由不同的处理图像自动产生,这样只要人为设置 α_θ 一个参数。

在用 PCNN 进行图像处理时,首先将一个二维 PCNN 网络的 $N_1 \times N_2$ 个神经元分别与二维输入图像的 $N_1 \times N_2$ 个像素相对应,相邻神经元存在连接,每个像素点的灰度值输入到对应神经元的连接输入,同时每个神经元的输出与其邻域中其他神经元输入相连。在 PCNN 迭代处理过程中,阈值随时间指数衰减,当某一神经元的内部活动项大于阈值时点火激活,被激活的神经元则通过与之相邻神经元的连接作用激励邻接神经元,若邻接神经元的内部活动项大于阈值,则被捕获激活,此时其输出为 1,否则输出为 0。显然,如果邻接神经元与前一次迭代激活的神经元所对应的像素具有相似强度,则邻接神经元容易被捕获激活,反之则不能被捕获激活。

3 滤波算法描述

3.1 PCNN 噪声点检测

噪声检测的目的是尽可能把噪声点从图像中找出来,为算法后续噪声滤除提供可靠的保证,一种良好的噪声检测方法应尽量避免漏检或误检的发生。由于脉冲噪声是叠加在原图像上的,其灰度值与其周围像素之间有很大的差异性,即噪声点不是高亮点 I_{ij}^{max}(或接近高亮点)就是低暗点 I_{ij}^{min}(或接近高亮点),根据脉冲噪声特点和 PCNN 相似神经元同步点火发放脉冲的特性,可快速有效地检测出噪点来,具体如下所述。

首先把含噪图像 I_{ij}(也可表示成 $I(i,j)$)输入 PCNN 网络,运用式(1)~(6)经过首次 PCNN 运行使灰度为 I_{ij}^{max} 的像素先点火激活,再进行第二次 PCNN 迭代处理会自动地把介于 $(I_{ij}^{max}/(1+\beta_{ij}L_{ij}), I_{ij}^{max})$ 之间的像素捕获激活,使两次输出 $Y_{ij} = 1$;其次对原含噪图像进行反白变换(最小值最大化变换),即 $I_{ij} = I^{max} - I_{ij}$ 处理,然后对再 I_{ij} 按前述 PCNN 方法进行两次迭代处理,并使其输出 $Y_{ij} = 1$。

这样经过 PCNN 处理后,输出 $Y_{ij} = 0$ 对应原图像像素点为信号点,而 $Y_{ij} = 1$ 对应的像素点有三种可能:(1)孤立的噪声点;(2)非孤立噪声点;(3)图像本身的信号像素点,需进一步对 Y 进行处理区分噪点。此时运用 3×3 元素全为 1 的模板 M 对 PCNN 输出 Y 做滑动处理,并判定以 $Y_{ij} = 1$ 为中心 M 邻域中为 1 的元素个数 N_Y:(1)当 $1 \leq N_Y < 3$ 时判定为孤立噪声点;(2)当 $3 \leq N_Y \leq 8$ 时判定为非孤立噪声;(3)当 $N_Y = 9$ 时判定为信号像素点,并使其对应的 $Y_{ij} = 0$。

3.2 第一级自适应滤波

通过上述 PCNN 噪点检测,为兼顾图像噪声滤除和细节保护两方面的性能,自适应选择滤波窗口及处理方法,当 M 中为 1 的个数 N_Y 较小(噪声密度较低)可选取小尺寸的滤波窗口以加强对图像细节的保护作用;当 N_Y 较大(噪声密度较高)可选取大尺寸的滤波窗口以增强图像去噪能力,据此,选取方形滤波窗口 H,

并可按下式来确定窗口大小 $h \times h$：

$$h = \begin{cases} 3, & 1 \leq N_Y \leq 2 \\ 5, & 3 \leq N_Y \leq 6 \\ 7, & 7 \leq N_Y \leq 8 \\ other operation, & N_Y = 9 \end{cases} \quad (7)$$

这样可以对含噪图像进行从左到右从上到下的自适应窗口选择滤波。

(1) 当 $Y_{ij} = 0$ 时，直接输出原图像中对应像素的灰度值，即 $f(i,j) = I(i,j)$

(2) 当 $Y_{ij} = 1$ 且 $N_Y = 1 \sim 8$ 时，可采用上述不同大小的滤波窗口 $\boldsymbol{H}$，实现对 $I(i,j)$ 像素的滤波，其数学表达描述如式(7)所示：

$$f(i,j) = \sum_{r,s \in H_{ij}} u(r,s) I(r,s) \quad (8)$$

式(7)中 $u(r,s)$ 是滤波窗口内的加权系数，其描述如下：

$$u(r,s) = \begin{cases} \dfrac{1/\{1 + \max[Q_{ij}, (I(r,s) - P_{ij})^4]\}}{\sum_{r,s \in H_{ij}} 1/\{1 + \max[Q_{ij}, (I(r,s) - P_{ij})^4]\}}, & Y(r,s) = 1 \\ 0, & Y(r,s) = 0 \end{cases} \quad (9)$$

$$Q_{ij} = \frac{\sum_{r,s \in H_{ij}} (I(r,s) - P_{ij})^4}{N_Y}, Y(r,s) = 1 \quad (10)$$

式中 P_{ij} 为滤波窗口中所有信号像素的灰度中值。

3.3 第二级多方向信息中值滤波

第一级自适应滤波在很大程度上滤除了噪声的干扰，但是当图像受到严重脉冲噪声污染时，无法达到较满意的滤波效果。为了对第一级滤噪后的图像在尽可能保持边缘及细节等信息的前提下进一步滤除残留噪声影响，引入进行第二级多方向信息中值滤波，其具体的滤波算法可描述如下：

$$M_1(i,j) = med[f(i,j+k); -K \leq k \leq K] \quad (11)$$

$$M_2(i,j) = med[f(i+k,j+k); -K \leq k \leq K] \quad (12)$$

$$M_3(i,j) = med[f(i+k,j); -K \leq k \leq K] \quad (13)$$

$$M_4(i,j) = med[f(i+k,j-k); -K \leq k \leq K] \quad (14)$$

$$M_5(i,j) = med[f(i \pm 1,j), f(i,j), f(i,j \pm 1)] \quad (15)$$

$$M_6(i,j) = med[f(i \pm 1,j \pm 1), f(i,j)] \quad (16)$$

式中 $f(i,j)$ 为第一级滤波后图像点 (i,j) 处的灰度值，$2K+1$ 为滤波模板大小，$M_n(i,j)(n=1,2\cdots,6)$ 表示运用不同模板时相应邻域的中值。

$$Z_{\min}(i,j) = \min[M_1(i,j), M_2(i,j), M_3(i,j), M_4(i,j), M_5(i,j), M_6(i,j)] \quad (17)$$

$$Z_{\max}(i,j) = \max[M_1(i,j), M_2(i,j), M_3(i,j), M_4(i,j), M_5(i,j), M_6(i,j)] \quad (18)$$

则多方向信息中值滤波的输出为:

$$Z_{out}(i,j) = med[Z_{\min}(i,j), Z_{\max}(i,j), f(i,j)] \quad (19)$$

多方向信息中值滤波在去噪的同时较好地保护图像的边缘和细节信息,特别适合作为与其他滤波方法相结合的二次细微辅助滤波。图1所示为本文噪声检测及滤波算法总流程图。

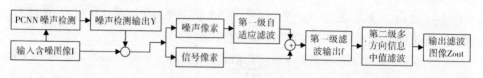

图1 算法流程框图

4 实验结果与分析

为验证本文所提算法(PM)的有效性,分别对256×256,灰度级为256的Lena、Cameraman和Rice等图像在受10%~90%不同噪声密度污染后的图像进行本文算法实验测试,并与传统中值滤波(TM)、多级中值滤波(MM)、文献[16]提出的PCNN赋时矩阵滤波法(PTM)及文献[11]提出的局部空间像素自适应滤波法(LPA)作了比较;对滤波性能的客观衡量用峰值信噪比(PSNR)、信噪比改善因子(SIF)和均方误差(MSE)等指标来描述;实验中参数$\alpha_\theta = 0.09$,W为3×3元素为1的矩阵,$K = 2$。

图2中(a1)~(c1),(a2)~(c2),……,(a7)~(c7)列图像分别为Rice、Lena和Cameraman原始图像,受50%、60%和70%噪声污染图像以及分别采用5种不同滤波算法滤波结果图像;图3、图4及图5是对Lena图像在受10%~90%脉冲噪声污染时,分别采用五种滤波算法的PSNR、SIF以及MSE(为作图比较方便,采用ln(MSE))随噪声密度变化的关系曲线;表1为三幅图像在受80%脉冲噪声污染时,分别采用五种滤波算法的三种测试指标PSNR、SIF以及MSE的对比实验数据。

通过实验仿真,从上述实验图表可以得出:本文提出的算法对受低密度还是对受高密度脉冲噪声污染的图像、或是对纹理细节简单还是纹理细节复杂的不同图像,都具有良好的滤波去噪性能,去噪后图像的主观视觉效果和客观评价指标PSNR、SIF及MSE均好于TM、MM、PTM及LPA等方法的滤噪结果;充分说明了在噪声检测阶段改进型PCNN较好地利用其神经元同步脉冲发放特性比较精确地区分和定位噪声点和信号点,为后续滤波奠定了良好基础,同时也反映了其后两级去噪过程中第一级自适应滤波和第二级多方向信息中值滤波的有机结合,在较好地保护图像边缘与细节的前提下圆满完成了去噪任务;且由实验图表可见随着噪声密度的增加,所提算法的视觉效果和客观指标评价一直保持较好的优越性,更进一步证明了本文算法在滤波去噪方面,特别是滤除受严重脉冲噪声污染图像的噪声方面更加表现了良好的主观效果,较强的去噪性能、优良的抗畸变能力、较稳定的鲁棒性和适应性。

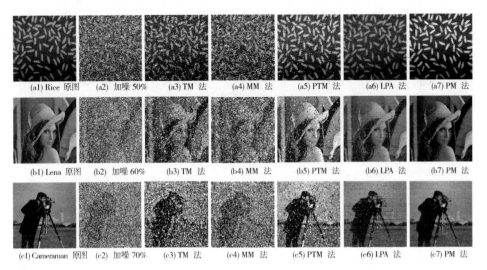

图2 对三幅不同脉冲噪声污染图像采用不同方法的滤波结果

表1 对3幅含噪80%图像采用几种不同滤波方法的
PSNR、SIF 及 MSE 对比实验数据

指标	Lena			Cameraman			Rice		
	PSNR	SIF	MSE	PSNR	SIF	MSE	PSNR	SIF	MSE
加噪图像	6.7944	0	137100	6.2561	0	15519	6.7525	0	13843
TM	8.5156	1.7020	9224.1	7.8873	1.6216	10660	8.4264	1.6854	9415.4
MM	7.0264	0.2128	12997	6.4679	0.2023	14780	6.9369	0.1959	13268
PTM	10.0815	3.5427	6431.7	9.5408	3.4876	7284.4	10.2988	3.8133	6117.8
LPA	17.0380	10.2436	1296.2	15.0124	8.7563	2066.5	17.0513	10.2987	1292.3
PM	20.4443	12.8588	707.41	17.2424	10.9896	1236.6	19.3657	12.6090	758.421

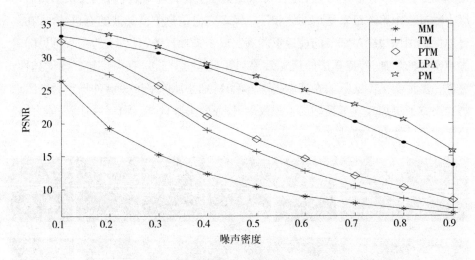

图 3 Lena 图像去噪时不同算法的 PSNR 随噪声密度变化比较

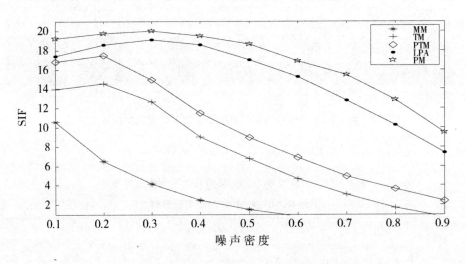

图 4 Lena 图像去噪时不同算法的 SIF 随噪声密度变化比较

5 结论

利用 PCNN 具有相似神经元同步激发脉冲的性质和脉冲噪声点灰度与其相邻信号像素灰度之间存在明显差异的特点进行噪点检测,再根据噪声点局部邻域信息对噪声进行第一级自适应滤波,然后再对其结果运用多方向信息中值二级细微辅助滤波。仿真结果表明,该算法在精确定位噪点的基础上,有效地滤除了脉冲噪声，同时，较好地保护了图像边缘细节信息，去噪后的图像具有更好地主观视觉效果和客观评价指标，比传统中值滤波及本文提到其它算法具有更好的滤波性能，且在噪声检测中无需设定检测阈值，其信噪比高、去噪能力强和适应性好，特别是对受严重噪声污染的图像，该算法显示了更大的优越性。但是本文算法与传统中值滤波等算法相比由于追求了滤波性能，所以算法在运行时间的开销上有所加长，对实时图像处理会产生一定的影响。

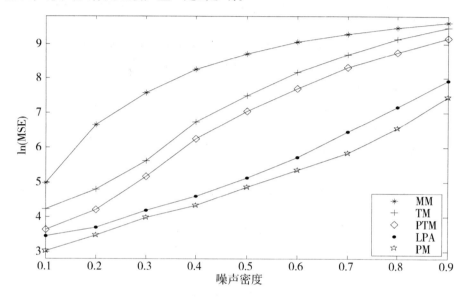

图5　Lena图像去噪时不同算法的ln(MSE)随噪声密度变化比较

参考文献

［1］Huang T S,Yang G J,Tang G Y. Fast two – dimensional median filtering algorithm［J］. IEEE Trans Acoustics,Speech,Signal Process,1979,ASSP – 1(1):13 – 18.

［2］Krishnan Nallaperumal,Justin Varghese et al. Selective Switching Median Filter for the Removal of Salt & Pepper Impulse Noise［C］. 2006 IFIPInternationalConference on Wireless and Optical Communications Networks ［A］. Bangalore,Indian,2006.

［3］Huang Zan. A Median Filter Based on Judging Impulse Noise by Statistic and Adaptive Threshold［C］. 2008

InternationalCongress on Image and Signal Processing［A］. Sanya,Hainan,China,2008,207 – 210.

［4］Pei – Eng Ng,Kai – Kuang Ma. A Switching Median Filter With Boundary criminative Noise Detection for Extremely Corrupted Images［J］. IEEE Trans on Image Processing,2006,15

(6):1510-1516

[5]邢藏菊,王守觉,邓浩江,等.一种基于极值中值的新型滤波算法[J].中国图象图形学报,2001,6(6):533-536.

XING Cang-ju,WANG Shou-jue,DENG Hao-jian. et al. A New Filtering Algorithm Based on Extremum and Median Value[J]. Journal of Image and Graphics,2001,6(6):533-536.

[6]Chung-Chia Kang. Wen-June Wang. A Novel Directional Median Filter Based on Fuzzy Reasoning[C].2008 IEEE International Symposium on Consumer Electronics[A]. Algarve, portugal,2008,1-4

[7]Xiaowei Han,Junsheng Li,Yanping Li,et al. A Selective and Adaptive Image Filtering Approach Based on Impulse Noise Detection[C]. Proceedings of the 5th World Congress on Intelligent Control and Automation[A]. Hangzhou. P. R. China,2004,4156~4159.

[8]Thou-Ho(Chao-Ho)Chen,Chin-Pao Tsai,Tsong-Yi Chen. An Intelligent Impulse Noise Detection Method by Adaptive Subband-Based Multi-State Median Filtering[C]. 2007 Second International Conference on Innovative Computing, Information and Control[A]. Kumamoto,Japan,2007,236~239.

[9]王瑄,毕秀丽,马建峰.基于二次噪声检测和细节保护规则函数的图像滤波算法[J].电子学报,2008,36(2):381-385.

WANG Xuan,BI Xiu-li,MA Jian-feng. Image Filtering Algorithm Using Double Noise Detector and Edge-Preserving Regularization Function[J]. ACTA ELECTRONICA SINIC,2008,36(2):381-385.

[10]G R Arce,R E Roster. Multilevel median filter:properties and efficacy[C]. Proc. ICASSP [A]. NewYork,1988,824-827.

[11]常瑞娜,穆晓敏,杨守义等.基于局部空间像素特征的自适应加权滤波算法[J].计算机工程与应用,2008,44(6):45-47.

CHANG Rui-na,MU Xiao-min,YANG Shou-yi. et al. Adaptive weighted filtering algorithm based on the feature of local spatial pixels[J]. Computer Engineering and Applications,2008,44(6):45-47.

[12]Johnson J L,Padgett M L. PCNN models and applications [J]. IEEE,Trans,Neural Networks,1999,10(3):480-498.

[13]Ma Y D,Dai R L,Li L. Image segmentation of embryonic plant cell using pulse-coupled neural networks [J]. Chinese Science Bulletin. 2002,47(2):167-172.

[14]刘勍,马义德,钱志柏.一种基于交叉熵的改进型PCNN图像自动分割新方法[J].中国图象图形学报,2005,10(5):579-584.

LIU Qing,Ma Yi-de,Qian Zhi-bai. Automated image segmentation using improved PCNN model based on cross-entropy [J]. Journal of Image and Graphics,2005,10(5):579-584.

[15]LIU Qing,MA Yi-de,ZHANG Shao-gang,et al. Image Target Recognition Using Pulse

Coupled Neural Networks Time Matrix[C]. Proceedings of the 26th Chinese Control Conference[A]. Zhangjiajie,Hunan,China. 2007,7:96 – 99.

[16]刘勍,马义德. 一种基于PCNN赋时矩阵的图像去噪新算法[J]. 电子与信息学报,2008,30(8):1869 – 1873.

LIU Qing,MA Yi – de. A New algorithm for Noise Reducing of Image Based on PCNN Time Matrix[J]. Journal of Electronics & Information Technology,2008,30(8):1869 – 1873.

注:本文曾发表在2009年11月《光电子·激光》第11期上。

天水古民居的建筑艺术与文化内涵研究

南喜涛[*]

本文简述了天水古城形成的历史,在调研的基础上对天水古民居进行分类,并介绍了天水典型的名人宅院,对天水古民居的建筑艺术与文化内涵进行了理论概括和总结。

天水历史悠久,约在公元前 11 世纪,天水地域即出现了先秦西犬丘,西犬丘即后来的西垂,是秦族、秦文化的发祥地,今天水市秦州区西南一带的礼县。西晋太康三年(282)置秦州,七年复置,并将秦州州治及天水郡治俱由冀城(今甘肃甘谷县)迁至上邽城,从此开始,上邽城即今天水城一直作为州、郡治所而成为陇东南地区的政治、经济、文化、军事中心。

一、天水古城概况

天水是一座文化底蕴深厚、建城历史久远、独具特色的国家级历史文化名城。公元 290 年前,今天水城就成为陇东南地区的军事、交通、经济、政治文化中心。

唐时,天水城成为丝绸之路上的繁华之地。公元 759 年,杜甫流寓秦州时,有"降虏兼千帐,居民有万家"的诗句,是唐肃宗首开用茶帛向少数民族换马的大型茶马市场。

明清时期的天水城是今天水城的前身(图 1 乾隆《直隶秦州新志》州城图),前后 600 多年间,天水城形成了"五城"格局,这种格局的五城,是东西一字排开,形如串珠,城城相连。康熙二十六年(1687 年),城内设 32 里,有长安里、西厢里、西宁里、玉泉里、仁义里、中和里、坊下里等。

---

[*] 作者简介:南喜涛(1962—),男,甘肃通渭人,天水师范学院教授、学士,主要从事古民居建筑结构研究。

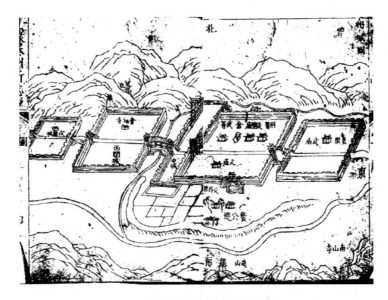

图1 乾隆《直隶秦州新志》州城图

民国二十八年(1939年)时(图2 天水市1939年城区略图),从东到西的带形城中尚有十余条大街和近100条主要巷道。而遍布大街小巷的民居院落是古城的主要组成部分。

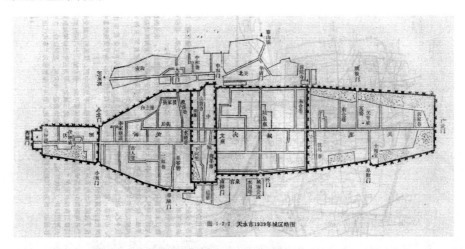

图2 天水市1939年城区略图

二、天水古民居四合院类型和典型的名人宅院

根据我们的调研统计,天水古城现存的古民居四合院有140处左右,分布在

历史上形成的东西五城之内,而现存的古民居院落大多数分布在西关城和伏羲城内。

(一)四合院形制与类型

天水四合院以两进为多,也有三进、四进者,两进之间多由过厅划分为两个庭院。这种四合院由大门、门道、小天井、照壁、垂花门、庭院、上房(主房)、左右厢房、下房(倒座)、耳房、月拱门等基本元素组成。其大门一般设在宅院左下角,进入大门后经过门道和小天井,左转九十度,穿过二门(垂花门),才进入正院或序院,从而使得庭院幽静、安全,私密感很强。

1. 基本型:即只有一进院的四合院,如澄源巷张庆麟故居北院,飞将巷9号院等。

2. 串联型:各庭院由一条纵轴线贯穿相联,以二进院和三进院居多。串联型又可分为两类:一类是隔垂花门式,如澄源巷张庆麟故居南侧两大院、伏羲路140号等。另一类是隔过厅式,属于这种形式的院落较多,如石作瑞故居、任士言故居、自由路14号、杨家楼19号等等。

3. 并联型:左、右院落组群并排相接组成的院落群,如自由路14号是并联式院落的典型。

4. 并排型:如甘肃省文物保护单位石作瑞宅院西侧,有四个基本型宅院并排组合而成。

5. 混合型,如澄源巷丁特故居。

(二)天水典型的名人宅院

1. 胡氏民居(南北宅子)

明代建筑群,国家重点文物保护单位,位于天水市秦州区市区中心街道民主路南北两侧。其中南宅子(图3 胡氏民居之南宅子)建于明嘉靖年间至隆庆年间,为明嘉靖丙午(1546年)秦州举人、中宪大夫、山西按察副使胡来缙的私宅。北宅子建于明万历四十三年(公元1615年),为胡来缙之子胡忻及其后代的居所,胡忻为进士出身,初任山西临汾知县,以后升为工部给事,官至太常寺少卿署

图3 胡氏民居之南宅子

正卿。

2. 石作瑞宅院

清代建筑群,甘肃省文物保护单位,位于天水市秦州区解放路石家巷,为清通仪大夫四川道台石作瑞的私宅(图4 石作瑞宅院主房)。

3. 哈锐宅院

哈锐(1862—1932),是一名回族翰林,祖籍福建,康熙初年西迁天水。哈锐故居位于今天水市秦州区市区澄源巷13号、17号,串联式一进三院,南北纵轴,大门东南角,前堂后寝,主厅房为两层,并有后院,保存完整,是清代建筑群,为甘肃省文物保护单位(图5 哈锐宅院主楼)。

图4 石作瑞宅院主房

图5 哈锐宅院主楼

4. 张庆麟宅院

张庆麟生于清嘉庆二十三年(即1818年),清咸丰庚申科考取进士后,又于同治壬戌科补行殿试,后被派往河北广平县任知县。张庆麟宅院位于天水市秦州区市区澄源巷西北角,是由21号、23号、42号院落组成的清代建筑群,为甘肃省文物保护单位(图6 张庆麟宅院主楼)。

5. 连腾霄宅院

连腾霄为清代秦州知府太学生,连腾霄宅院位于天水市秦州区北关连家巷,该建筑群为一进三院的清代建筑群,是甘肃省文物

图6 张庆麟宅院主楼

保护单位(图7 连腾霄宅院主楼)。

图7　连腾霄宅院主楼

6.杨名显宅院(杨家宅子)

杨家宅子坐落在秦州区西关杨家楼65号、67号,属于典型的明代建筑,系清初通奉大夫、广东承宣布政使司布政使杨名显祖居(图8 杨名显宅院)。

图8　杨名显宅院

除了上面所列举的一些典型的名人宅院之外,还有诸如飞将巷古民居建筑群;共和巷的冯国瑞宅院;三星巷的赵家祠堂、光绪年间武进士赵子培的故居、原国家副主席荣毅仁的居所和荣氏当年复兴面粉厂、共产国际人士葛霁云故居;澄源巷周氏家族古建筑群;砚房背后贾家公馆;自治巷张氏三院等。

三、天水古民居的建造特点

(一)外观简朴优美

天水古民居与青砖黛瓦的山西、北京等北方民居不同,房屋外墙一般用土砖填充砌筑,很少夯土作,一些房子山墙顶部的三角形部位才略砌青砖,以利防水。因此,天水明清民居几乎全为木结构,青瓦土墙,质地朴实。背向院落之外的墙面一般不施粉刷,简单朴素。

庭院四周外围的房屋一般为单庇向各自院内延伸,因而造成后墙高峻、深巷高墙的景观。具有凝重浓郁的地方性特点和原发性的古朴。

天水古民居的屋面同其他类型古建筑屋面一样,往往都是因屋架举折形成的凹曲面,这种"反宇飞檐"的造型,使负重的屋顶变得十分轻盈。这种弧面很有利于雨水的排放,既保护了檐下的墙体,又有视觉上丰富的流线美感。

(二)内向含蓄之美

天水历史上处于封闭的内陆腹地,民风朴实而具有自己的特点,民居建筑亦然。院落外各立面展现的是封闭的后檐墙外观,很少开窗,整座院落对外只有一道大门开出,显得厚实冷峻、封闭严实。而院落内四面的单体建筑为庭院空间提供了丰富的观赏与实用界面,这种界面一般处于金柱位置,且大面积都是柔美的木质门窗,这种浓郁的内向界面强化了庭院的内向品格。

天水古民居院落,一般在四面都有廊道,这种廊道空间,很好地体现了中庭式院落由公共空间(庭院)到"模糊空间"(廊道)再到私密空间(室内)的空间层次关系,这种层次关系使廊道既有明显的室外内化性,同时又体现着室内外化性,通过这个中介性的"过度空间",使得室内空间与庭院空间内外渗透,将大自然的天空、阳光、山水、花木,纳入生活环境之中,富有娴静恬淡、闲情逸致之趣。这种庭院既是生活空间,又是休闲佳地,把生活和休闲统一在一个总体空间之中,取得景观与环境两相宜的艺术效果。如上文所列举的这些名人宅院等,在这些院落中,给人一种强烈的内向含蓄,人与自然和谐共存的境界。

(三)布局中的伦理秩序意象

天水古民居是典型的中国传统建筑的院落式布局,表现出了严整纵深的庭院组合,中轴突出的对称格局。房屋布置遵循前次后主、中轴为上、东西为次的等级观念,主次分明。相对高大的厅堂、楼居多布置在主轴线上,并加以重点装修。空间划分与功能分区远近有分,内外有别。内眷多居内宅内院,外眷则居外宅外院,是地位、尊严、安全的象征,这种空间处理手法与传统"前殿后寝"同出一辙。通过正院与偏院,正房与厢房,正房与倒座,外院与内院,前庭与后庭等等空间的主从、内外划分,使庭院组群充分适应了封建礼教严格区分尊卑、上下、亲疏、贵贱、男女、长幼等伦理秩序需要,体现出儒家"礼"文化意象。

如南北宅子、贾家公馆、自由路 12 号、自由路 14 号、自治巷张氏三院、哈锐宅院、张庆麟宅院、连腾霄宅院、冯国瑞宅院等都是这种布局的代表。

(四)突出重点装修

相对高大的厅堂、楼房多布置在主轴线上,并加以重点装修(如前文图 5 哈锐宅院主楼)。

垂花门、砖雕影壁、窗棂等是天水古民居中重点装修的部位,是具有标志性的建筑艺术品。这些重点装修都使天水古民居比中原民居更具有一种本色的质朴、通透和轻巧、活泼。

天水民居多木雕,垂花门是天水民居的木雕精品。而其他单体建筑上的木雕集中在檐口下的装修上,由雀替、大小耍头、柱上铺做和柱间铺做以及铺做间花板组成。内檐装饰有碧纱橱、落地柜、炕柜等,均美观而不落繁琐花哨的俗套。

天水古民居的木雕不同于南方徽派民居和北方晋商民居雕刻的堆砌繁缛,其线条回转流畅,简洁粗犷,具有美观大方亦不失精雕细琢的雕刻风格(见图 9:麒麟、图 10:草龙)。

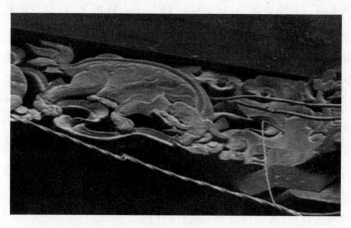

图 9　麒麟

图10 草龙

砖雕主要见于大门内天井的照壁上,为仿木砖雕,雕刻图案以佛八宝、暗八仙、五蝠捧寿、松竹梅岁寒三友等内容多见,既增添了民俗文化气氛又不显豪华奢靡。

(五)色彩本质

传统民居的色彩,从春秋时期起,不断发展,大致到了宋朝已有定局。在春秋时代,多使用原色,经过长期的发展,在色彩的调和与对比方面积累了很多经验。南北朝、隋、唐时期的官邸、府第多用灰砖、红柱,并在柱、檐、枋、斗拱等上面绘有各种彩画。明朝的色彩已趋成熟完整。清代多注重鲜明艳丽,但随着府第等级的高低和各地区的不同,又有若干的差别。

北京民居四合院大多采用紫朱油或红土烟子油。这种紫色调的暖红色,可与其周围的青绿油画、大面积的青砖灰瓦产生冷暖对比,营造出一种亲切热烈的气氛。山东民宅的色彩大体可归纳为灰色、红柱、粉墙、棕色梁架及红、绿门窗,大门一般为黑漆红边,形成清秀淡雅的格调。

天水民居,其色彩则可总结为"自然本色"四个字。外观的灰调土墙为自然本色,内院的梁柱、门窗、家具,亦为自然木质本色,总体呈现出崇尚自然淡雅的"水墨画"式格调,不热衷于庙宇式的彩绘油饰。例外者则少数。

四、结语

天水地处中国的西北地区,古丝绸之路从长安出发,西入甘肃,途经的第一个重镇就是天水,这里东接陕西,南通四川,自古以来就是陇东南的交通要冲,深受南北东西各种风俗文化的综合影响。

天水古民居建筑墙体用材少砖石而多土墙,外观显得质朴敦厚。其院落总体布局主次分明,纵串横并,外封闭而内开敞,注重将人的主观情意和自然物境互相交融,即便是庭院内花木扶疏,池鱼叠石也充满着富贵吉祥与子孙繁衍家族永恒的寓意。单体房屋建筑的凹曲屋面呈反宇飞檐,造型十分优美;屋身则尺度适中,金柱位置的前檐墙大面积为木质纹理的门窗,檐口下铺做突出线条层次,注重木雕手法和雕刻题材；木雕垂花门和砖雕照壁，雕刻线条回转流畅，简洁粗犷，风格美观大方而不细密缛繁，总体色彩呈现出崇尚自然淡雅的"水墨画"式的自然本色。这些使得天水古民居既有北方古建筑的封闭厚重，又有江南古民居的秀丽灵巧，在北方建筑厚墙灰瓦的质感中，浸透着南方建筑的秀巧轻盈，体现出天水古民居建筑的独特风格。

参考文献

[1] 董鉴泓.中国城市建设史[M].北京:中国建筑工业出版社,1989.

[2] 天水市城乡建设保护委员会.天水城市建设志[M].兰州:甘肃人民出版社,1994.

[3] 南喜涛.天水古民居[M].兰州:甘肃人民出版社,2007.

[4] 雍际春、吴宏岐.陇上江南——天水[M].西安:三秦出版社,2003.

注:该认文原发表在2010年《小城镇建设》第6期上

后　记

六十年风雨历程,六十年求索奋进。编辑出版《天水师范学院60周年校庆文库》(以下简称《文库》),是校庆系列活动之"学术华章"的精彩之笔。《文库》的出版,对传承大学之道,弘扬学术精神,展示学校学科建设和科学研究取得的成就,彰显学术传统,砥砺后学奋进等都具有重要意义。

春风化雨育桃李,弦歌不辍谱华章。天水师范学院在60年办学历程中,涌现出了一大批默默无闻、淡泊名利、潜心教学科研的教师,他们奋战在教学科研一线,为社会培养了近10万计的人才,公开发表学术论文10000多篇(其中,SCI、EI、CSSCI源刊论文1000多篇),出版专著600多部,其中不乏经得起历史检验和学术史考量的成果。为此,搭乘60周年校庆的东风,科研管理处根据学校校庆的总体规划,策划出版了这套校庆《文库》。

最初,我们打算策划出版校庆《文库》,主要是面向校内学术成果丰硕、在甘肃省内外乃至国内外有较大影响的学者,将其代表性学术成果以专著的形式呈现。经讨论,我们也初步拟选了10位教师,请其撰写书稿。后因时间紧迫,入选学者也感到在短时期内很难拿出文稿。因此,我们调整了《文库》的编纂思路,由原来出版知名学者论著,改为征集校内教师具有学科代表性和学术影响力的论文分卷结集出版。《文库》之所以仅选定教授或具有博士学位副教授且已发表在SCI、EI或CSSCI源刊的论文(已退休教授入选论文未作发表期刊级别的限制),主要是基于出版篇幅的考虑。如果征集全校教师的论文,可能卷帙浩繁,短时间内难以出版。在此,请论文未被《文库》收录的老师谅解。

原定《文库》的分卷书名为"文学卷""史地卷""政法卷""商学卷""教育卷""体艺卷""生物卷""化学卷""数理卷""工程卷",后出版社建议,总名称用"天水师范学院60周年校庆文库",各分卷用反映收录论文内容的卷名。经编委会会议协商论证,分卷分别定为《现代性视域下的中国语言文学研究》《"一带一路"视域下的西北史地研究》《"一带一路"视域下的政治经济研究》《"一带一路"视域下的教师教育研究》《"一带一路"视域下的体育艺术研究》《生态文明视域下的生物学研究》《分子科学视域下的化学前沿问题研究》《现代科学思维视域下的数理问题研究》《新工科视域下的工程基础与应用研究》。由于收录论文来自不同学科领域、不同研究方向、不同作者,这些卷名不一定能准确反映所有论文的核心要义。但为出版策略计,还请相关论文作者体谅。

鉴于作者提交的论文质量较高,我们没有对内容做任何改动。但由于每本文集都有既定篇幅限制,我们对没有以学校为第一署名单位的论文和同一作者提交的多篇论文,在收录数量上做了限制。希望这些论文作者理解。

这套《文库》的出版得到了论文作者的积极响应,得到了学校领导的极大关怀,同时也得到了光明日报出版社的大力支持。在此,我们表示深切的感谢。《文库》论文征集、编校过程中,王弋博、王军、焦成瑾、贾来生、丁恒飞、杨红平、袁焜、刘晓斌、贾迎亮、付乔等老师做了大量的审校工作,以及刘勋、汪玉峰、赵玉祥、施海燕、杨婷、包文娟、吕婉灵等老师付出了大量心血,对他们的辛勤劳动和默默无闻的奉献致以崇高的敬意。

《天水师范学院60周年校庆文库》编委会

2019年8月